MW00838195

Germ
Unity

1970   1980   1990

Janecke, *Ion Exchange Polymers*

etter, *Electrochemical Kinetics*

Institute for Electron Microscopy (IFE)

uska, *Electron Microscopy*

Zeitler, *Experimental Electron Microscopy*

Schlögl, *Catalyst Development*

rill, *atalysis, -ray Analysis*

Block, *Surface Chemistry, Spectrometry*

Freund, *Modeling Catalytic Systems*

Gerischer, *Surface Electrochemistry*

Ertl, *Catalysis, Metal Clusters*

Wolf, *Elementary Processes at Surfaces*

Iolière, *Solid-State Electron Diffraction*

Scheffler, *Computational Physics and Chemistry*

Institute for Physical Chemistry

Bradshaw, *Adsorption, Surface Oxidation*

Meijer, *Molecular Physics*

orrmann, *X-ray rystal Optics*

Institute for Structure Research

"Leaving behind the obsolete is the true tradition of science."
Heinz Gerischer, 1969 (translated from German)

Hosemann, *Polymer, Colloid, and Crystal Structure*

berreiter, *hysical Chemistry of Polymers*

Blue: fundamental physical chemistry / chemical physics
Green: structure research
Red: applied research

James, Steinhauser, Hoffmann, Friedrich

One Hundred Years at the Intersection of Chemistry and Physics

Published under the auspices of the Board of Directors of the Fritz Haber Institute of the Max Planck Society:

Hans-Joachim Freund
Gerard Meijer
Matthias Scheffler
Robert Schlögl
Martin Wolf

Jeremiah James · Thomas Steinhauser ·
Dieter Hoffmann · Bretislav Friedrich

One Hundred Years at the Intersection of Chemistry and Physics

The Fritz Haber Institute of the Max Planck Society
1911–2011

De Gruyter

Authors:

Dr. Jeremiah James
Fritz Haber Institute of the
Max Planck Society
Faradayweg 4–6
14195 Berlin
james@fhi-berlin.mpg.de

Prof. Dr. Dieter Hoffmann
Max Planck Institute for the
History of Science
Boltzmannstr. 22
14195 Berlin
dh@mpiwg-berlin.mpg.de

Dr. Thomas Steinhauser
Fritz Haber Institute of the
Max Planck Society
Faradayweg 4–6
14195 Berlin
thomas@fhi-berlin.mpg.de

Prof. Dr. Bretislav Friedrich
Fritz Haber Institute of the
Max Planck Society
Faradayweg 4–6
14195 Berlin
brich@fhi-berlin.mpg.de

Cover images:
Front cover: Kaiser Wilhelm Institute for Physical Chemistry and Electrochemistry, 1913. From left to right, "factory" building, main building, director's villa, known today as Haber Villa.
Back cover: Campus of the Fritz Haber Institute of the Max Planck Society, 2011. The Institute's historic buildings, contiguous with the "Röntgenbau" on their right, house the Departments of Physical Chemistry and Molecular Physics. Below the "Röntgenbau" is the building of the infrared free electron laser. The top-most building on the right houses the Department of Inorganic Chemistry. The Institute's workshops are located in the two-towered building. The hexagonal structure houses the Theory Department and the Joint Network Center. The two buildings to the left on the lower side of the areal view house the Department of Chemical Physics. The red-roofed building next to Haber Villa is Willstätter House, which houses part of the Theory Department.

ISBN 978-3-11-023953-9
e-ISBN 978-3-11-023954-6

Library of Congress Cataloging-in-Publication Data

One hundred years at the intersection of chemistry and physics : the Fritz Haber Institute of the Max Planck Society, 1911–2011 / by Jeremiah James ... [et al.].
    p. cm.
  ISBN 978-3-11-023953-9
  1. Max-Planck-Gesellschaft zur Förderung der Wissenschaften. Fritz-Haber-Institut. 2. Electrochemistry–Research–Germany–History. 3. Physics–Research–Germany–History. 4. Haber, Fritz, 1868-1934. I. James, Jeremiah. II. Title: Fritz Haber Institute of the Max Planck Society, 1911–2011.
  QD558.2.G32M396   2011
  541′.37072043--dc23
                                          2011028402

Bibliografic Information published by the Deutsche Nationalbibliothek
The Deutsche Nationalbibliothek lists this publication in the Deutsche Nationalbibliografie; detailed bibliographic data are available in the Internet at http://dnb.d-nb.de.

© 2011 Walter de Gruyter GmbH & Co. KG, Berlin/Boston

Typesetting: PTP-Berlin Protago-TEX-Production, Berlin
Printing and binding: Hubert & Co. GmbH & Co. KG, Göttingen
♾ Printed on acid-free paper

Printed in Germany

www.degruyter.com

Contents

Contents

Preface

The Kaiser Wilhelm Institute for Physical Chemistry and Electrochemistry was established in 1911 as one of the first two institutes of the Kaiser Wilhelm Society (KWG). Its successor, the Fritz Haber Institute (FHI), is not only one of the oldest and most tradition rich institutes of the Max Planck Society (MPG), but also one of the most distinguished, with the highest number of affiliated Nobel Laureates of any KWG/MPG institute. These include Fritz Haber, the founding director, the later directors Max von Laue, Ernst Ruska and Gerhard Ertl, and several scientists who served at the Institute in lesser capacities, such as James Franck, Eugene Wigner and Heinrich Wieland.

The Institute has been not only a hub of scientific excellence and productivity but also an active participant in the history of the 20th century. It played a central role in German poison-gas research and the conduct of chemical warfare during World War I. It was particularly hard-hit by Nazi racial policies and was revamped into a "National Socialist Model Enterprise;" then to remain productive during the Cold War, it had to assert itself in a territorially insular and politically precarious West-Berlin.

In order to do justice to the complex scientific and political history of the FHI, the Institute's Board of Directors, prompted by the approaching centenary of the Institute (and of the KWG/MPG), offered support in 2007 for a broad historical investigation of the Institute from its inception to the present. The Centennial Group, established in response to the Board's initiative in the Fall of 2008 and comprised of the undersigned, launched a research project to examine in detail the changing relationships between this long-standing scientific Institute, its rapidly expanding scientific subject matter and the tumultuous political history of the past hundred years.

Although historians and social scientists alike have published several studies on the overarching Kaiser Wilhelm and Max Planck Societies, they have not lavished similar attention on the individual research institutes.[1] For the FHI in particular, there have been noteworthy, purpose-driven studies that have attempted to span the entire history of the Institute, but they remain quite brief and were not intended to present balanced historical accounts.[2] Certain KWG/MPG institutes have also garnered space in broader historical works, and the FHI is prominent among them. In these histories, however, the FHI is often a bit player in what are

---

1 Brocke, Laitko, *KWG Institute*.
2 Chmiel, Hansmann, Krauß, Lehmann, Mehrtens, Ranke, Smandek, Sorg, Swoboda, Wurzenrainer, *Bemerkungen*; MPG, *FHI I*. New edition: MPG, *FHI II*.

primarily biographies of famous scientists such a Fritz Haber,[3] Michael Polanyi,[4] Peter Adolf Thiessen[5] and Robert Havemann.[6] Or, since the Institute was so closely coupled to social and political events, it appears as a prominent part of monographs focused on topics such as the founding of the KWG,[7] poison gas research[8] and Nazi era science.[9] Although detailed and well-founded, the sum of these studies fails to provide a balanced history of the Fritz Haber Institute. Still wanting was an historical study of the Institute, supported by archival research, that presented a long-term view of the Institute, and hence could more adequately address the rapid and sustained changes in the intellectual content of the sciences to which it contributed and in the societies, both scientific and political, that supported it.

The founding of the KWG amounted to the third in a series of institutional innovations – after the founding of the Berlin University (1810) and of the Imperial Institute of Physics and Technology (1887) – which originated in Berlin and helped shape the modern research system. In a sense, the founding of the Kaiser Wilhelm Institute for Physical Chemistry and Electrochemistry can be regarded as one of the consequences of the Prussian "Althoff System," credited with the modernization of education and research structures in Germany. It came about in reaction to forewarnings by numerous prominent scientists and science-policy makers about the waning of Germany's scientific and technological superiority relative to the US and to other European nations. In hindsight, the founding of the KWG in general and of the KWI for Physical Chemistry and Electrochemistry in particular could be viewed as a successful answer to this challenge, for during the following decades the KWG established itself nationally and internationally as a leading research organization. Although the creation of the KWG broke new ground for the state funding of science in Germany, the establishment of the KWI for Physical Chemistry and Electrochemistry was made possible by an endowment from the Berlin Banker and philanthropist Leopold Koppel, granted on the condition that Fritz Haber, well-known for his discovery of a method to synthesize ammonia from its elements, be made the institute's director.

As indicated above, the history of the Institute has largely paralleled that of 20th-century Germany. It undertook controversial weapons research during World War I, followed by a "Golden Era" during the 1920s and early 1930s, in spite of financial hardships. Under the National Socialists it experienced a purge of its scientific staff and a diversion of its research into the service of the new regime, accompanied by a breakdown in its international relations. In the immediate aftermath of World War II it suffered crippling material losses, from which it recovered

---

3 Szöllösi-Janze, *Haber*; Stoltzenberg, *Haber*.
4 Nye, *Polanyi*.
5 Eibl, *Thiessen*.
6 Hoffmann, *Havemann*.
7 Johnson, *Chemists*. Wendel, *Kaiser-Wilhelm-Gesellschaft*.
8 Groehler, *Tod*; L.F. Haber, *Poison*; Schmaltz, *Kampfstoff-Forschung*.
9 Deichmann, *Flüchten*; Hachtmann, *Wissensmanagement*.

Table 1. *Nobel Laureates affiliated with the KWI for Physical Chemistry and Electrochemistry or the Fritz Haber Institute of the Max Planck Society.*

	Year of award and whereabouts	Nobel-Prizewinning work done at	Period at KWI–PChE/ FHI–MPG	Capacity
Max von Laue (1879-1960)	1914, Munich (LMU)	Munich (LMU)	1951–1959	Director
Fritz Haber (1868–1934)	1918, Berlin (PChE)	Karlsruhe (THK)	1911–1933	Founding Director
James Franck (1882–1964)	1924, Göttingen (GAU)	Berlin (FWU)	1918–1920	Department Leader
Heinrich Wieland (1877–1957)	1927, Munich (LMU)	Freiburg (ALU), Munich (LMU)	1917–1918	Fellow, Army Officer
Eugene Wigner (1902–1995)	1963, Princeton	Berlin (PChE, THCh), Princeton	1923–1932	PhD Student, Fellow
Ernst Ruska (1906–1988)	1986, Berlin (FHI)	Berlin (TUB, Siemens, FHI)	1949–1974	Director of the IFE
Gerhard Ertl (*1936)	2007, Berlin (FHI)	Munich (LMU), Berlin (FHI)	1986–2004	Director

ALU	Albrecht-Ludwig-Universität Freiburg
FWU/HU	Friedrich-Wilhelms-Universität/Humboldt-Universität zu Berlin
GAU	Georg-August-Universität Göttingen
IFE	Institut für Elektronenmikroskopie am Fritz-Haber-Institut der MPG
KWI-PChE/FHI-MPG	KWI für Physikalische Chemie und Elektrochemie/Fritz-Haber-Institut der MPG
LMU	Ludwig-Maximilans-Universität München
THCh/TUB	Technische Hochschule Charlottenburg/Technische Universität Berlin
THK	Technische Hochschule Karlsruhe

slowly in the post-war era. In 1952, the Institute took the name of its founding director and, in 1953, joined the fledgling Max Planck Society, successor to the Kaiser Wilhelm Society. During the 1950s and 1960s, the Institute supported diverse researches into the structure of matter and electron microscopy. In subsequent decades, as both Berlin and the Max Planck Society underwent significant changes, the institute reorganized around a board of coequal scientific directors and a renewed focus on the investigation of elementary processes on surfaces and interfaces, topics of research that had been central to the work of Fritz Haber and the first "Golden Era" of the Institute but that had never before been developed into an institute-wide research orientation.

The shifting fortunes and socio-political roles of the Institute help to explain the striking breadth of topics that have been researched within its walls over the past century, but so too do the diverse abilities and personalities of the scientists who have made the Institute, however briefly, their intellectual home. Dozens of distinguished scientists, among them the already mentioned seven Nobel laureates, have shaped the pace-setting research in physical chemistry, chemical physics and related fields performed at the Institute. Their interests have ranged from providing for the concrete needs of society, in times of peace or war, to plumbing the abstract depths of quantum mechanics, and from the apparent simplicity of hydrogen chemistry to the acknowledged complexity of non-linear dynamics. Their investigations reflect a distinct, intellectual facet of 20th-century history which is inextricable from social, cultural and political history.

Over the three years of its existence, the Centennial Project has worked toward three goals. The first and foremost has been to produce this volume, which spans the history of the FHI and is based largely on as yet untapped archival material. Laboring against a deadline set one hundred years ago, its authors have striven to bridge the institutional and scientific history of the Institute and to provide a holistic picture up to the present. Second, the Centennial Project has nurtured more detailed and rigorous studies on specific themes, aimed at engaging the history of science community. Finally, the Centennial Group reached out to the wider public by putting on twenty seminars which revolved around the key figures and themes in the history of the FHI both as part of the research necessary for the historical overview and in order to provide a forum for broader collaborations among scholars already interested in aspects of the history of the FHI.

In our efforts we have been frequently reminded of the words of a doyen of modern history of science research, Gerald Holton:[10]

> [T]he science research project of today is the temporary culmination of a very long, hard-fought struggle by a largely invisible community of our ancestors. Each of us may be standing on the shoulders of giants; more often we stand on the graves of our predecessors.

At times in the history of the Fritz Haber Institute, these struggles have been more than "simply" intellectual and have, in themselves or through their outcomes, had profound and even fatal, repercussions. The Centennial Project – and this volume – has aimed to highlight these struggles of the past and to pay tribute to those who, for the most part, persevered through them. We hope that the historical perspective offered herein improves understanding of the Institute's place within the educational and research establishments and helps to raise historical awareness amongst scholars working at the Institute and beyond.

Berlin, June 2011

Bretislav Friedrich
Dieter Hoffmann
Jeremiah James
Thomas Steinhauser

---

10   Gerald Holton, *Pais Prize Lecture*.

Acknowledgments

Before descending into the past, we would like to take the opportunity to thank all those who have helped make the present book possible.

First and foremost, we would like to thank the Board of Directors of the Fritz Haber Institute, who, in preparation for the 100th anniversary of the founding of the Institute, initiated the Centennial Project, amongst whose goals was the production of this volume. The Institute and its directors Hans-Joachim Freund, Gerard Meijer, Matthias Scheffler, Robert Schlögl and Martin Wolf have generously supported the Centennial Project over the last three years and followed with an abiding interest the writing of this book. The directors and their coworkers also provided us with extensive materials relating to scientific activities at the Institute, especially during the recent past, and much of Chapter 6 is based on their generosity. Our thanks go also to the administrative director Karsten Horn for his dedicated support.

Several other members of the FHI staff also deserve our thanks for their enthusiastic support and ongoing assistance throughout the project. We owe our gratitude to: Katrin Quetting and Uta Siebeky of the FHI Library for their help in locating resources at the FHI and beyond; Bärbel Lehmann for allowing us access to her private collection of photographs of the FHI and directing us toward several of the images used in this book; Waruno Mahdi from the Department of Physical Chemistry and Albrecht Preusser from the MPG Joint Network Center for providing assistance in preparing the photographs and figures for publication.

As part of the institutional and interdisciplinary cooperation between the FHI and the MPI for the History of Science, the Centennial Project overlapped significantly with the project on the History and Foundations of Quantum Physics (HFQP). Special thanks are therefore also due to the MPI for the History of Science and its director Jürgen Renn, who, along with his colleagues involved in the HFQP project, helped us to resolve several tangled questions concerning the history of the FHI through discussions, planned and spontaneous. In addition, the Institute offered us essentially unlimited access to its expertise and resources relating to the history of science in general, and its library provided us with outstanding reference services.

A number of affiliates of the FHI and the MPG also helped us gain deeper insights into the workings of the Institute in the post-WWII era through interviews, informal conversations, and comments on our work. Prominent among them were: Hans Bradaczek (Berlin), Alexander Bradshaw (Berlin), Ruth Broser and Immanuel Broser (Berlin), Manuel Cardona (Stuttgart), Gerhard Ertl (Berlin), Utz Havemann–von Trotha (Ferch), Hellmut Karge (Berlin), Reimar Lüst (Hamburg),

Ellen Reuber (Berlin), Joachim Sauer (Berlin), Manfred Swoboda (Berlin), Klaus Thiessen (Neuenhagen), Knut Urban (Jülich), Harald Warrikhoff (Berlin), Burkhard Wende (Berlin) and Elmar Zeitler (Berlin). We are indebted to all of them for their assistance.

Our thanks are also due to Eckart Henning (Berlin), Hubert Laitko (Berlin), Inga Meiser (Berlin), Falk Müller (Frankfurt/Main), Gabor Pallo (Budapest), Michael Schaaf (Johannesburg), Mary Jo Nye (Corvallis), Jeffrey Johnson (Villanova), Phil Bunker (Ottawa) and Florian Schmaltz (Frankfurt/Main) for valuable suggestions and for discussions of specific topics.

History of science in general, and institutional histories in particular, rely heavily on archival research, and we would like to express our gratitude to all of the archives and archival staff that have supported our work on this project. Special thanks are due the Archive of the Max Planck Society in Berlin. Its director Lorenz Beck and staff members Bernd Hoffmann, Joachim Japp, Marion Kazemi, Susanne Uebele and Dirk Ullmann, helped us access not only the catalogued material related to the FHI but also untapped sources that greatly enriched this history. In this connection we would also like to express our gratitude to Werner Hofmann, former chairman of the Chemical-Physical-Technical Section of the Max Planck Society, for granting us access to the Section's records.

We are also grateful to the speakers who participated in the Centennial Seminar Series. Their in-depth knowledge of specific aspects of the history of the FHI and related topics both expanded our own horizons and enlightened their audiences.

Two student assistants provided invaluable service to the Centennial Project, Felix Ameseder (Technische Universität Berlin) and Hannah Riniker (Humboldt Universität zu Berlin). Our sincerest thanks to both of them.

Last but not least, we would like to thank the de Gruyter Publishing House Berlin, and in particular Alexander Grossmann, who embraced the idea of a book about the one-hundred year history of the Fritz Haber Institute and was willing to undertake the production of both a German and an English version. Katrin Nagel, Simone Schneider and Ulrike Swientek from the editorial department also provided valuable advice on the layout of the book, as well as kind and patient guidance during the lengthy process of bringing this volume to press.

Without the multifaceted help of those acknowledged above, and many others who supported us and our work, it would hardly have been possible to finish this book in the limited time available to us. This aid and assistance notwithstanding, the material included in this volume has been selected by the authors alone and presented in the manner we felt appropriate. We alone are answerable for the interpretations of historical events it offers, as well as any lacunas or inaccuracies that may have escaped our notice.

1 "under my protection and name...." – Origins and Founding of the Institute

> Today there are entire disciplines that simply no longer fit within the bounds of the colleges and universities, either because they require such extensive machinery and instrumentation that no university department can afford them, or because they concern problems that are too advanced for students and can only be tackled by junior scholars.[1]

So wrote the Berlin theologian and director of the Royal Library, Adolf Harnack, in a 1909 memorandum that would serve as the founding document for the Kaiser Wilhelm Society (*Kaiser-Wilhelm-Gesellschaft*, KWG) and hence for the Kaiser Wilhelm Institute for Physical Chemistry and Electrochemistry (*Kaiser-Wilhelm-Institut für physikalische Chemie und Elektrochemie*, KWI), one of the first institutes established by the society. The creation of these organizations was the culmination of interwoven chains of events stretching back well into the 19th century and closely tied to the rise of Germany, and Berlin in particular, as an international center for scholarly research. Three institutional innovations contributed substantially to this rise to academic prominence.[2] The first was the founding in Berlin in 1810 of the Friedrich Wilhelm University (*Friedrich-Wilhelms-Universität*, Berlin University), one of the first establishments to promote the ideal of the unity of research and teaching, which would become a hallmark of the modern research university. Then in 1887, the Imperial Institute of Physics and Technology (*Physikalisch-Technische Reichsanstalt*, PTR) began operations in Berlin-Charlottenburg. The first large research institute to stand outside the university system, the Imperial Institute of Physics and Technology, resulted from a close collaboration between the state, industry and science, aimed at establishing an institute that could meet the demands of modern, large-scale scientific research. Finally, came the founding of the Kaiser Wilhelm Society for the Promotion of the Sciences in 1911 – the last in this series of institutional innovations that, though initiated in Berlin, would affect the scientific landscape well beyond the borders of Germany. The Kaiser Wilhelm Society was established to supplement the efforts of the universities and technical colleges in the natural sciences and engineering, in part as a response to rising international competition, particularly from the United States, whose rapid scientific growth had already begun to call German leadership in these fields into question.

---

1 Harnack, *Denkschrift*, p. 82.
2 Laitko, *Innovationen*.

Fig. 1.1. *The Imperial Institute of Physics and Technology in Berlin-Charlottenburg.*

The founders of the Kaiser Wilhelm Society built upon the successes of the Imperial Institute of Physics and Technology. Researchers in other scientific disciplines, chemistry in particular, sought to emulate the model it provided of an institute for "big science," in the modern sense, dedicated exclusively to physics and metrology.[3] The Berlin chemist, Emil Fischer, who had founded a laboratory for quantitative research in chemistry based on precision methods just before the turn of the century, became the spokesperson for the effort to create an Imperial Institute of Chemistry analogous to the Institute of Physics and Technology. Initial attempts to establish such an institute through government means foundered on issues of state finances. In response, leading representatives of academic and industrial chemistry established in 1905 an independent planning committee to promote the prospective institute, which then developed into an Imperial Institute of Chemistry Association some three years later.[4] The association aimed principally to collect the funds necessary for the construction and maintenance of the proposed institute through donations and membership dues, and thereby circumvent dependence upon state financing, although association members remained dedicated, on principle, to state support for the new institute. A memorandum on the need for an Imperial Institute of Chemistry composed by Emil Fischer in

3 Cf. Cahan, *Institute*.
4 Cf. Johnson, *Chemists*.

collaboration with the renowned physical chemists Wilhelm Ostwald and Walther Nernst provided the immediate impetus and the occasion for the formation of the Association. In the memo, Fischer, Nernst and Ostwald laid particular weight upon the promotion of physical chemistry, arguing that it should constitute the "scientific backbone" and the central division of the new institute.

All this occurred against the backdrop of a rapid boom in classical, organic synthetic chemistry during the last third of the 19th century in Germany, which formed the basis for the production of ever more complex synthetic dyes and supported the associated large chemical concerns, but which left behind such sub-disciplines as inorganic chemistry and the young and aspiring fields of biochemistry, physiological chemistry and physical chemistry. Institutional support for physical chemistry was particularly meager. The field initially crystallized around only a handful of organizational structures in Germany: Wilhelm Ostwald's institute in Leipzig, the associated research school and the newly established Journal of Physical Chemistry (*Zeitschrift für physikalische Chemie*).[5] The founding of Ostwald's Institute at Leipzig University in 1887 was not part of a great wave of new institutes for physical chemistry. At the beginning of the 20th century there remained only a few, relatively small institutes and some subaltern posts dedicated to the field, although these could provide excellent research facilities, as was the case for Walther Nernst in Göttingen and for Fritz Haber in Karlsruhe. The shortfall in Berlin was particularly striking; only the *extraordinarius* professors Hans Landoldt and Hans Jahn represented the field, which hardly sufficed for the promotion of the capital city to a peak research position.[6]

This lack of institutional support appeared an ever more acute problem in that physical chemistry was no longer a liminal field, but was increasingly recognized as a fundamental discipline within chemistry. Physical chemists wanted to address basic concepts common to all of chemistry, such as chemical binding and chemical reactions, which touched upon the underlying physical bases of chemical structure. The resulting, increasingly multi-faceted new branch of chemistry could not be neatly inserted into the German ordinary professoriate, which was still marked by stubborn disciplinary boundaries and the almost overwhelming dominance of organic chemistry. Hence, there was an enormous demand for new institutes of physical chemistry, and not only for small, specialized institutes that could make up for the existing deficit but also for a central institute, preferably housed in Berlin, the imperial capital, that could help guide the development of the field. With respect to its size, facilities and modernity the Chemical Institute of the Berlin University erected in 1900 for Emil Fischer presented an excellent model for such a flagship institute[7] – assuming, of course, one overlooked its focus on organic chemistry.

---

5 Girnus, *Grundzüge*.
6 Bartelt, *Berlin*.
7 Reinhardt, *Zentrale*.

Although the chemical industry offered "substantial donations" to support the proposed institute, and representatives of the Prussian state expressed no doubts concerning the significance of such an undertaking, the lack of government funds continued to block progress, and the state set aside discussion of the matter early in 1909.[8] But soon thereafter a new opportunity to promote the project arose in connection with the grand designs of the preeminent director of academic affairs in the Prussian Ministry of Culture, Friedrich Althoff. As the new century opened, Althoff had formulated a plan to develop the remaining crown lands in the former demesne of Dahlem into "a German Oxford." In Althoff's vision, the Berlin suburb would host not only annexes of the Berlin University but also new research institutes and extensive scientific collections.[9] However, Althoff died in 1908, without having made significant progress toward realizing his plans. Nevertheless, shortly after Althoff's death, Kaiser Wilhelm II commissioned Althoff's long-time associate Friedrich Schmidt (after 1920 Schmidt-Ott) to compile a report on "Althoff's plans for Dahlem." Less than a year later, as the Prussian bureaucracy began to ponder an appropriate royal gift for the centennial of the Berlin University, Schmidt-Ott sent Althoff's plans to the Chief of the Civil Cabinet, Rudolf von Valentini, who then forwarded them to the theologian and Director of the Royal Library, Adolf von Harnack, along with his own request for a report on the plans. As part of his report, Harnack was supposed to evaluate the present scientific standing of Germany and to develop from his assessment a proposal for a fitting centenary gift from the Kaiser. Harnack completed his report, a "Memorandum concerning the founding of a Kaiser Wilhelm Institute for scientific research," in the autumn of 1909.

In his memorandum, Harnack relied not only upon Althoff's plans but also upon the recommendations of scientists such as Emil Fischer, Walther Nernst and August Wassermann, weaving these together with a dire warning concerning the plight of German science and the concomitant dangers to state and business interests:

> ...German scholarship lags behind that of other nations in important lines of scientific research and its ability to compete is gravely threatened...This circumstance is already ominous for the nation-state and is becoming ever more so for scholarship. For the state, it is ominous because in these times of extraordinarily intensified nationalist sentiment, unlike in the past, every result of scholarly research is stamped with a national seal. [10]

Harnack paid particular attention to the problems of theoretical and organic chemistry. He emphasized the importance of research on the chemical elements and atomic weights, which he lauded as:

---

8 Cf. Johnson, *Chemists*, p. 48 ff.
9 Cf. Engel, *Dahlem*.
10 Harnack, *Denkschrift*, p. 82.

Fig. 1.2. *Adolf Harnack (1851–1930) in the official robes of the President of the Kaiser Wilhelm Society.*

a science set apart. Every advance in this field is of great import for the entirety of chemistry, but this discipline no longer fits within the framework of higher education. It demands its own laboratories.[11]

Harnack also highlighted the precarious situation of organic chemistry, which was struggling with an ongoing movement of advanced research from institutes of higher education to industrial laboratories. In light of the special significance of chemistry to German science and industry and the fact that "significant preparations had already been made," Harnack recommended that his patrons "begin with the founding of a large chemical research institute," with other research institutes to follow later.[12]

That Harnack's memorandum focused so clearly on chemistry was due in no small part to his having sought advice on the project from Emil Fischer and others who had taken part in the activities of the Imperial Institute of Chemistry Association. Harnack wrote in his memorandum, contrary to its later realization, of a *single* grand chemical research institute to be supported not by the state alone but by "a cooperation of the state and wealthy, scientifically-interested, private citizens." To achieve this, "an association of donors would be established that stretched across the entire monarchy," and hence possessed the financial resources necessary to realize such grand plans.

Harnack's memorandum was read to the monarch, word for word, at the beginning of December, and received the "liveliest, unrestrained applause of his Majesty."

---

11 Ibid.
12 Ibid., p. 87.

Fig. 1.3. *Festivities celebrating the 100th anniversary of Berlin's Friedrich Wilhelm University on 11 October 1910.*

The memo also formed the basis of the speech Wilhelm II delivered at the ostentatious centenary festivities of the Berlin University, on 11 October 1910. In the new, grand auditorium of the university the Kaiser announced his plan:

> to found under [his] protection and name a society tasked with the establishment and maintenance of new research institutes...establishments that go beyond the framework of the institutions of higher education and serve only research, uninfluenced by instructional goals, although in close contact with the academies and universities.[13]

Furthermore, the Kaiser could report in good conscience to the assembled guests that, in addition to having received "hearty declarations of approval" for his plan, he had received pledges that would amount to a sizable endowment for the new society, on the order of 10 million marks.[14]

To raise funds for the endowment, state executives worked through presidents of regional councils and city mayors, sometimes with the help of representatives of the finance ministry, to identify those citizens with the largest fortunes and thereby establish a pool of prospective donors. The mayor of Düsseldorf at the

13 MPG, *50 Jahre KWG/MPG*, p. 113.
14 Ibid., p. 114.

time spoke of "surrounding the noble quarry for Professor Harnack and his Royal Protector."[15] These "noble quarry" were then invited to make sizable donations, both as a sign of their sympathy for communal needs and royal interests and as a way to serve the national cause of scholarship. The strategy was successful and pledges grew rapidly, in spite of limited support from the landed nobility and other representatives of the old elite, as prominent members of the aspiring industrial and banking bourgeoisie came forward to offer their support. Among the most generous donors ranked the steel magnate Gustav Krupp von Bohlen und Halbach, as well as the directors of leading technical firms Wilhelm von Siemens of Siemens Electrical and Henry Theodore von Böttinger of Bayer Chemicals, and the bankers Franz von Mendelssohn and Eduard Arnhold. Donors with Jewish ancestry were conspicuously overrepresented amongst these key contributors. In a manner reminiscent of the "court Jews" of an earlier era, many of these Jewish benefactors sought social recognition, as well as concrete political and financial opportunities that might otherwise be barred to them by anti-Semitism, through their involvement in a grand national project.

A few short weeks after the Kaiser's proclamation, on 11 January 1911, seventy-nine donors assembled in Berlin under the chairmanship of the Prussian Minister of Culture, August von Trott zu Solz, for the constitutive meeting of the Kaiser Wilhelm Society. They decided questions of institutional structure, drafted a constitution and appointed the first ten senators of the Society. The privilege of appointing the remaining ten senators belonged to the Kaiser, ensuring the social exclusivity of the chief administrative organ of the Society. At its first official session, two weeks later, the senate predictably elected Adolf von Harnack President of the Society, a function he initially fulfilled voluntarily and in addition to his existing duties to the state.

The Kaiser Wilhelm Society was clearly a private research organization, supported by a private endowment and registered in Berlin as a private association, but the state had no intention of absenting itself from the administration of the Society. In his announcement at the Berlin University centenary, Wilhelm II forthrightly declared that "it will be the responsibility of [his] administration ... that the institutes to be founded ... do not want for state aid." This state aid took the form of a commitment to pay the operating expenses of the new institutes, as well as the salaries of the scholars they employed. Thus the KWG, though a private research organization, clearly came under the aegis of the state, through which the Society sought, in the words of its first president, to avoid "the risk of dependence upon clique and capital"[16] and to ensure the future of scholarly research against unforeseen vicissitudes. That said, the basic structure of the Society gave large donors considerable leverage, including allowing them a voice in the design of specific institutes, and some of the largest donations to the Society came with strict conditions, such as a gift from the Imperial Institute of Chemistry

---

15 Burchardt, *Wissenschaftspolitik*, p. 54.
16 Harnack to von Trott zu Solz, 22 January 1910, in: MPG, *50 Jahre KWG/MPG*, p. 95.

Fig. 1.4. *Cartoon from Simplicissimus: "The three magi bring their Christmas gifts".*

Fritz Haber (1868–1934)

Haber's name serves as an apt reminder of the Janus-face of modern science. On one side the industrial process of ammonia synthesis, developed by Haber jointly with Carl Bosch and Alwin Mittasch, is the basis for large-scale production of fertilizers and, as such, has greatly contributed to maintaining the food supply for the growing world population. On the other side, the Haber-Bosch process is also the basis for the mass production of explosives  and munitions. Moreover, Haber's research not only supported and, indeed, enabled Germany's prolonged involvement in WWI, but, during its course, Haber also became the 'father of chemical warfare' by directing his Kaiser Wilhelm Institute toward the development of poison gases, regarded by some as the first weapons of mass destruction. Whereas, in Haber's view, chemical weapons were supposed to break the stalemate of trench warfare (and preclude the slaughter of millions by artillery shells) by forcing the adversary to surrender.

Born into a Jewish family in Prussian Breslau, Haber studied chemistry in Berlin, graduating in 1891. After a string of minor industrial and university posts, he settled in 1894 at the Karlsruhe Technical University, where he received his habilitation and in 1898 became *extraordinarius* and in 1906 *ordinarius* (full) Professor of Physical Chemistry and Electrochemistry. Later he would refer to his time in Karlsruhe as "the best working years of my life." During the seventeen years in Karslruhe he not only laid the scientific foundations for the Haber-Bosch process, for which he would receive the 1918 Nobel Prize in Chemistry, but also became a well-known protagonist of physical chemistry through his contributions to the thermodynamics of gas-phase reactions; the scope and depth of this work led him to conclusions resembling the Third Law of Thermodynamics.

Haber's previous achievements clearly qualified him to become the founding director, in 1911, of the KWI for Physical Chemistry and Electrochemistry, which he would develop, especially during the Weimar era, into a world-renowned center of research at the intersection of chemistry and physics. His own research interests lay in reaction kinetics, as well in developing an electrochemical procedure for extracting gold from seawater. During the Weimar years he would become one of the most influential spokespersons for the Kaiser Wilhelm Society, dedicated in particular to repairing relations with the estranged international scientific community and to establishing the Emergency Association of German Science, the forerunner of today's German Research Foundation (DFG). In contrast to many of his colleagues, Haber embraced the Weimar Republic and ranked among its open supporters.

Still, neither his great scientific merits nor his unbridled patriotism sufficed to stave off his loss of status and position once the Nazis rose to power. Ill and heart-broken, Haber died in Basel less than a year after being driven from Germany. The strength of his bond with Germany is illustrated by the fact that, as late as 1933, he donated his ammonia synthesis apparatus to the Deutsches Museum in Munich, "the Walhalla of German science and technology."

role the personal tensions between Nernst and Haber played in these events is beyond the scope of this volume, but these tensions were clearly more pronounced than usual, even for colleagues competing for a prestigious post.[23]

Haber's official appointment as director of the newly-founded Kaiser Wilhelm Institute for Physical Chemistry and Electrochemistry and his release from his post in Karlsruhe would not be finalized until June of 1911; nevertheless, he rapidly became engrossed in the construction of the Berlin institute. In the months preceding his permanent move to Dahlem, Haber commuted frequently between Karlsruhe, where he still had teaching duties, and Berlin. While in Berlin, Haber not only offered input on architectural plans for the new Institute but also had a hand in drafting its charter and made suggestions concerning its future operation. So far as the charter was concerned, Haber aimed to ensure, in his own words:

> ...that the influence of the administrative organs and the advisory board did not exceed an acceptable level. Personally, [he had] grave misgivings concerning the cre-ation of arrangements that enabled a body assembled from disciplinary colleagues to affect the operation of an institute.[24]

The resulting charter established first and foremost the predominance of the Insti-tute Director, anticipating the so-called "Harnack Principle," which would become one of the guiding principles of the Kaiser Wilhelm Society. This principle stip-ulated that "the Society chooses an (outstanding) scholar and builds an institute around him [sic]."[25] The charter of Haber's institute read such that the director had sole authority to appoint scientific coworkers and accept guest researchers, independent of any representatives of the Koppel Foundation, donors to the Kaiser Wilhelm Society or relevant political authorities. Furthermore, the charter gave the director broad authority to decide questions concerning the use of the Institute's endowment and the installation of apparatus, and it declared him

> completely free in the exercise of his scientific activities, within the limits set by the budget; above all, he was subject to no restrictions concerning his choice and pursuit of scientific projects.[26]

In monetary matters the director was subject to the Koppel Foundation Council, which was the legal governing body of the Institute. In addition to the Institute Director and the Foundation Council, the charter established a Scientific Advisory Board, whose twelve members served five-year terms and provided guidance and suggestions concerning ongoing research at the Institute. The membership of the Advisory Board was not limited to top-notch scientists; instead, it included rep-resentatives from a broad range of scholarly institutions, including the Prussian Academy of Sciences, leading German Universities and the Kaiser Wilhelm Soci-ety. In practice, however, the Advisory Board did little to steer the Institute and

---

23 Johnson, *Chemists*, p. 123.
24 F. Haber to R. Willstätter, Karlsruhe 23 June 1911. Werner, *Haber Willstätter*, p. 43.
25 Brocke, Laitko, *KWG Institute*, p. 130.
26 Szöllösi-Janze, *Haber*, p. 230.

Fig. 1.7. *Dahlem near the end of 1918; in the foreground the Kaiser Wilhelm Institutes for Chemistry (left) and Physical Chemistry and Electrochemistry (right); in the background the KWI for Biology, opened in 1915.*

maintained instead "an ornamental character," thanks in part to the predominance of the director.

One interesting feature of the Advisory Board was that it also oversaw the neighboring KWI for Chemistry. This reflected not only the fact that physical chemistry was originally envisioned as just one division of an even larger chemical institute but also the immediate proximity of the two institutes, as well as plans for their parallel construction and inauguration. Since the buildings of the Kaiser Wilhelm Society were to be "nobly appointed," according to the desires of its Imperial Protector, the man entrusted with the design of the new chemical institutes was none other than court architect Ernst von Ihne. Ihne established his reputation as a designer of scholarly edifices with his designs for the new Prussian Royal Library, now the State Library Unter den Linden, and for the Kaiser Friedrich Museum, now the Bode Museum.[27] But Ihne was entrusted only with the exterior design of the institutes and ensuring that they remained true to the Wilhelmine ambiance of Dahlem. The interior design and technical details of the two institutes fell to the building planner Max Guth. Guth had established a reputation of his own for the design of modern research centers with his plans for section one of the Chemical Institute of the Berlin University and for the Materials Testing Office in Lichterfelde. In addition to Ihne and Guth, Haber took an active role in the design of his new institute, as advocate for the prospective needs of its scientific users. In this capacity, Haber made good use of his contacts at BASF. Before the end of 1910 Haber had already presented members of the firm responsible for

---

27 Jenrich, *Ihne.*

Fig. 1.8.   *On the left is the KWI for Chemistry, on the right the KWI for Physical Chemistry and Electrochemistry: to the extreme left and right are the Director's Villas.*

laboratory design with a 23 page description of the material needs of a modern research institute, as he saw them.[28] Once the experts at BASF had pointed out and improved upon the weak points in Haber's proposal the technical plans for the institute were essentially complete. Roughly simultaneously, at the beginning of 1911, the architects were instructed to begin construction of the Institute, taking into full consideration the proposals from Haber and BASF. The pains taken to achieve a symbiosis between aesthetic-architectural concerns and the demands of scientific research were embodied in features such as the gray façade, chosen "so that absolutely no colored light, which might disturb investigations, would enter the working rooms."[29] The institute encompassed a total volume of roughly 18,000 m^3 with 2,500 m^2 of usable floor space divided between a main building and a "factory building," which was dominated by a 200 m^2 "machine hall" that housed large apparatus and offered facilities adequate for the construction of small pilot works. These two facilities were connected by a 20 meter long enclosed walkway. A director's villa would also occupy the Institute's estate but was not habitable until 1913, a year after the inauguration of the Institute. The generous working space and top-notch research facilities of the Institute were intended to suffice not only Haber and his colleagues but also the numerous scientific guests anticipated in the charter of the Institute. In the factory building, chemical laboratories and technicians' workshops surrounded the central machine hall. The main building housed the "scientific division" of the Institute spread across two floors. In the basement were concentrated dark rooms and constant temperature rooms. The ground floor contained the director's laboratory, a calibration room, various workshops and a seminar room that seated approximately 25. Above these lay the library and the glassblower's workshop as well as laboratories for the department heads and scientific coworkers. On the top floor were instruments for photochemistry, and chemical and mineralogical sample collections. Both the

28   Szöllösi-Janze, *Haber*, p. 226.
29   MPG, *FHI*, p. 10.

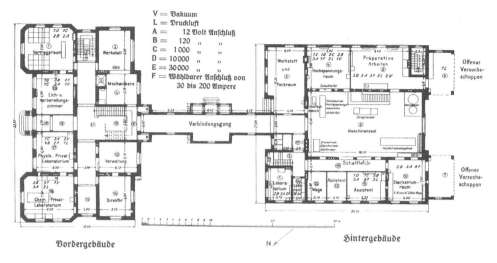

Fig. 1.9. *Ground floor of the KWI for Physical Chemistry and Electrochemistry, 1912.*

main building and the factory building were also furnished with small apartments for assistants and guests.

Overall the Institute was outfitted with more modern and high-performance equipment than most universities at the time. In addition to providing for standard chemical procedures, the laboratory furnishings enabled researchers to pursue a number of specialized lines of research – such as studies of thermodynamic constants in the constant temperature rooms. The apparatus at the Institute conformed to the latest standards in physical chemistry research. For example, the electrical equipment was the best available at the time, offering not only alternating and direct current facilities but also a high-voltage installation in the machine hall. Similarly, the Institute possessed apparatus for research at the extremes of available gas pressures and at the limits of spectroscopic precision.[30]

All of this took shape in the space of just over a year. The Prussian Minister of Public Works granted the building permit in May of 1911, so that construction could begin in the summer. One year later, in July of 1912, the framing and roofing were finished and work could begin on the interior of the facility. Before this work could proceed, however, Haber and the contractors had to solve the commonplace problem of having exceeded their construction budget. When it became clear that the initial endowment of 700,000 marks for construction and apparatus simply would not suffice, Koppel proved himself once more a magnanimous patron and promised an additional 300,000 marks for the project. His generosity would be

---

30 Description of the planned furnishings of the Institute, no title, Abt. Va, Rep. 5, Nr. 1789.

Fig. 1.10.   *The Kaiser underway to the inauguration of the Kaiser Wilhelm Institutes on 23 October 1912; behind him Adolf Harnack, Emil Fischer and Fritz Haber.*

rewarded with, among other things, an exemption from the standard gift tax and a personal introduction to the Kaiser. He was also the only donor mentioned by name in the Kaiser's address at the inauguration of the Institute.

The inauguration was an ostentatious affair celebrated on 23 October 1912. Johnson pithily characterized it as the outcome and expression of an,

> incident [that] epitomized the Prussian style of modernization through the Kaiser Wilhelm Society; the gleam of Koppel's gold had combined with the aristocratic aura of the Kaiser's person to produce the special brilliance of a new institute.[31]

It marked the opening of both the Institute for Physical and Electrochemistry and the adjacent Institute for Chemistry. Participants assembled in the library of the Chemistry Institute. The program was "at his Majesty's request, as restricted [in scope] as possible." It included brief addresses from Emil Fischer, Adolf Harnack, Culture Minister August Trott zu Solz and, of course, his royal Majesty.[32] Then came a tour of the Institutes with brief scientific talks and demonstrations. The presentations in the Physical Chemistry and Electrochemistry Institute were supervised by the Director himself, and among other things, included a demonstration of ammonia synthesis, which was presented as a practical application of fundamental chemical principles. At the conclusion of the celebration members gathered in the machine hall of the Physical Chemistry and Electrochemistry Institute for the first general assembly of the Kaiser Wilhelm Society.

31   Johnson, *Chemists*, p. 139.
32   MPG, *50 Jahre KWG/MPG*, p. 150–155.

This occurred almost a year to the day after the official founding of the Institute on 28 October 1911. This was the date on which the Board of the Koppel Foundation signed the legal documents establishing the Kaiser Wilhelm Institute for Physical Chemistry and Electrochemistry, in which the Foundation pledged not only to fund the construction and equipping of the new institute but also to contribute 35,000 marks annually for the next ten years to help cover its operating expenses. The remainder of the operating expenses and maintenance costs, including the salary of the director, fell to the Prussian state. This financial arrangement would differ from those of the other Kaiser Wilhelm Institutes, which were funded directly from the general accounts of the Kaiser Wilhelm Society. Moreover, the Institute for Physical Chemistry and Electrochemistry was formally under the direction of the Board of the Koppel Foundation not the Board of the Kaiser Wilhelm Society. However, a separate Advisory Board, established exclusively to help run the Institute, held the real executive authority and safeguarded the independence of the Institute. On the first such Advisory Board sat the Institute Director, Fritz Haber, as a non-voting member; Leopold Koppel, who was Chairman of the Board; Rudolf von Valentini, the Head of the Imperial Civil Cabinet, and Friedrich Schmidt-Ott and Councilman Klotz, both of whom represented the Prussian Ministry of Culture. Hence, though the Institute was named as part of the Kaiser Wilhelm Society, it was not administered by the Society and appeared separately in the Annual Reports of the Society up to 1923, at which point the annuity from the Koppel Foundation expired and financial hardships caused by staggering inflation necessitated the full integration of the Institute into the Kaiser Wilhelm Society.

Research in the name of the Institute also commenced in the autumn of 1911. As they lacked laboratories of their own in Berlin, Haber and his colleagues initially pursued their research as guests at various Berlin research centers, especially the Imperial Institute of Physics and Technology in Charlottenburg. As mentioned earlier, Haber was accompanied in his move to Dahlem by his Karlsruhe colleagues Richard Leiser and Gerhard Just, who were guaranteed positions at the new Institute. The Japanese expert on specific heat measurements, Setsuro Tamaru, also moved with Haber from Karlsruhe to Dahlem; he joined Leiser and Just as one of only three paid scientific coworkers of the Institute. The Institute also housed three unpaid scientific coworkers (Richard Becker, A. von Bubnoff, and William Ramsay, Jr. of London) as well as mechanics, laboratory assistants, secretaries and groundskeepers. There were also scientists who performed research at the Institute whose official positions remain somewhat unclear, such as the Finnish student Yrgö Kauko and Haber's former assistant Friedrich Epstein, both of whom followed Haber from Karlsruhe. In fiscal year 1912/13, after the move into the new institute buildings and the commencement of normal operations, the tally of scientific personnel rapidly climbed to five paid and thirteen unpaid coworkers, with approximately ten support staff aiding them. This remained the distribution of personnel until the outbreak of the First World War, which would lead to an extensive but tightly focused expansion of the Institute.

Fig. 1.11.   *Fritz Haber with his colleagues (left to right) Herbert Freundlich, Setsuro Tamaru and Reginald Oliver Herzog, circa 1913.*

Haber initially guided research at the Institute more according to his personal predilections than an explicit research program or even a central guiding principle. So far as these predilections were concerned, Haber admitted:

> When it comes to my scientific work, I would say: I have worked on things from several fields, but always jumped back and forth.[33]

The first scientific activities undertaken at the Institute were culminations of projects begun by Haber and his colleagues in Karlsruhe. Just continued research begun with Haber into electron emission in the course of gas-metal reactions, while Fritz Hiller, Haber's former PhD student, followed his mentor's lead in research on the inner cone of hydrocarbon flames. Haber also reaffirmed his interest in electrochemistry with an investigation of the effects on electrochemical reactions involving gases of passing currents through the walls of the gas containers. Nevertheless, the bulk of Haber's scientific publications in the years leading up to the First World War continued to relate to ammonia synthesis. In addition to an array of articles detailing new research on the subject, most of them co-written with Tamaru, Haber also published the results of researches undertaken with Robert LeRossignol, Haber's chief assistant during initial development of the synthesis process, and Harold Cecil Greenwood, both of whom had returned to Britain before Haber moved to Dahlem. These articles concerned, primarily,

---

33   Werner, *Haber Willstätter*, p. 55.

Fig. 1.12. *Fritz Haber and Richard Leiser with the firedamp whistle, 1913.*

the thermodynamics of the synthesis reaction and measurements of the specific heat of ammonia. In them Haber broadened his initially applications-oriented perspective on ammonia synthesis, reexamining the reaction in light of fundamental questions in physical chemistry.

Similarly central to research at the Institute in its early years was the development of a firedamp (methane) detector for use in coal mines.[34] Up to that point, the safety lamp developed by Sir Humphry Davy at the beginning of the 19th century was the preferred safety and warning apparatus. However, the lamps themselves posed something of a risk in that a defective lamp could set off an explosion. The Kaiser himself witnessed the effects of one such disaster during a visit to Krupp at Villa Hügel in Essen in the summer of 1912, and he used the opportunity of the opening of the Kaiser Wilhelm Institutes to request that German chemists develop a safer, more reliable detector. Haber had been informed in advance of the Kaiser's interest in such a device, and at the inauguration, he presented a gas interferometer he had developed in collaboration with the Zeiss company. However, the interferometer was a precision measurement device, not a rugged methane detector; hence, the practical problem remained unsolved.

---

34 Cf. Szöllösi-Janze, *Haber*, p. 237 ff.

Together with his assistant Richard Leiser, Haber dedicated himself during the next year to fulfilling the Kaiser's request. In so doing, he was entering into competition with colleagues at numerous other chemical institutes, among them the director of the neighboring KWI for Chemistry, Ernst Beckmann. Unlike his colleagues, who based their designs primarily upon spectroscopy or analytic chemistry, Haber began with acoustics, specifically the fact that the tone of a whistle depends upon the speed of sound in the gas that fills it. Using this fact, Haber and Leiser developed a firedamp whistle whose pitch would change when filled with methane. The whistle was the first successful answer to the challenge posed by the Kaiser, and Haber demonstrated it the following year at the general assembly of the Kaiser Wilhelm Society, in the presence of His Majesty. Haber signed over the patent and production rights for the whistle to Koppel's Auergesellschaft, but the device never saw widespread distribution and use. The whistle required precision machining that made its production costly, and it was not robust enough to withstand long periods of uninterrupted use. Moreover, it could not be properly calibrated on site. But even though the device was neither a rousing practical success nor a generator of great profits, it did add to Haber's scientific reputation and his symbolic capital. Haber's interest in firedamp detectors also illustrates the extent to which he oriented his research around the technical problems of his times – a feature of his scientific activities that would take on particular significance during the coming war.

Before the war, however, Fritz Haber and his new Institute would not only leave their mark on modern physical chemistry, but also contribute to the development of quantum theory. After 1911 and the first Solvay Conference in Brussels, quantum theory, which had up to that point been accepted by only a few scientists and remained largely in the shadows, began its move to the center of research interests in the physical sciences.[35] Haber was among those who expressed a deepening interest in the theory at the time, and he embarked upon an investigation of the relationship between quantum theory and chemical thermodynamics; the results of which scientists would discard, however, in light of subsequent research. In collaboration with Gerhard Just, Haber also pursued a series of researches on the emission of electrons during chemical reactions, specifically reactions between alkali metals and halide gases. Haber and Just drew parallels between this phenomenon and the photoelectric effect, for which Einstein had provided a quantum theoretical explanation some five years earlier, but Haber and Just were not able to produce a similarly succinct and robust explanation of the new electron emission phenomenon.

Haber's interest in quantum theory was also stimulated by his personal acquaintance with Albert Einstein; the two came to know and respect one another at a scientific meeting in Karlsruhe in the autumn of 1911.[36] Thereafter Haber joined a group of Berlin scholars who took great pains to bring Einstein to Berlin. By so

35 Hermann, *Frühgeschichte*, p. 140 ff.
36 Stern, *Freunde*.

doing, they aimed not only to decorate Berlin's scholarly society with the "rising star on the physics horizon"[37] but also to make use of his abilities in the fields of heat and radiation theory in order to develop a new quantum theory of matter and, thereby, promote the integration of physics and chemistry.[38] Haber in particular, hoped to expand the purview of physical chemistry and, hence, of his new Institute, through the admixture of quantum theoretical methods. In a letter from 1913, regarding the importance of attracting Einstein to Berlin, he explained:

> For me the decisive factor is that the development of theoretical chemistry, which, since the days of Helmholtz, has striven productively, under the leadership of van't Hoff, to make the achievements of heat theory its own, having effectively reached this goal, now strives further to enlist radiation theory and electrochemistry in service to its endeavors. This fundamental task can be promoted incomparably by the admission of Mr. Einstein to our circle of institutes.[39]

But Einstein did not fulfill these hopes. He concentrated instead on the completion of the general theory of relativity during his early years in Berlin.[40] Moreover, the installment of Einstein as a full-time member of the Berlin Academy obviated the original plan to tie Einstein directly to the Kaiser Wilhelm Society by offering him a position at Haber's Institute. Although Einstein did establish an office at the Institute shortly after his arrival in Berlin in 1914, it is unclear precisely how long he remained a guest in Dahlem. The arrangement would certainly have come to an end when the Institute moved under military auspices early in 1916, especially considering that Einstein had an apartment in central Berlin by that point.

Still Haber's efforts to combine physical chemistry with quantum theoretical methods was not limited to the person of Albert Einstein. In the Summer of 1913, the Breslau physical chemist Otto Sackur arrived at the Institute as a "scientific guest." When Richard Leiser accepted an industry position in the spring of 1914, Sackur then took over as department head. Sackur's pre-war research explored the borderlands between physical chemistry, statistical mechanics and quantum theory.[41] Sackur developed a quantum theory of the monatomic ideal gas that was pioneering in many respects, not least of all in the connections it forged between these three fields of research. The theory was part of a larger project to formulate a quantum theoretical description of the motion of atoms and molecules in gases. Up to that point, scientists had only quantized the motion of periodic systems, such as harmonic oscillators, radiating oscillators and lattice vibrations. Sackur's ideal gas theory made him one of the pioneers of the quantization of translational motion. Initially his colleagues were not sympathetic to his efforts. Walther Nernst, for example, fundamentally rejected the possibility of quantizing translational motion, and Arnold Eucken saw no need to do so because existing

37 Hoffmann, *Einsteins Berlin*.
38 Barkan, *Witches Sabbath*.
39 F. Haber to H. Krüss, Pontresina 4 November 1913, Einstein, *CPAE*, Bd. 3, pS.511; cf. also Haber, *Körper*, p. 117 ff.
40 Cf. Fölsing, *Einstein*, p. 414 ff.; Renn, *Einstein Kontexte*
41 Cf. Badino, Friedrich, *Sackur*.

Fig. 1.13.   *Albert Einstein and Fritz Haber in the stairwell of the KWI, circa 1914.*

experimental results offered no hints of related quantum phenomena. Sackur was not discouraged though. He wanted to use quantum methods to provide more satisfying solutions to long-standing problems in thermochemistry, including the calculation of the absolute entropy of gases, which in turn would allow more precise calculation of equilibrium constants. The Sackur-Tetrode equation, developed independently but very nearly simultaneously by Sackur and the Dutch physicist Hugo Tetrode, provided a quantum-statistical expression for the entropy of a monatomic ideal gas valid in the limit of high temperatures and low densities. While at the Institute, Sackur added an experimental component to his researches, attempting to detect deviations from the classical ideal gas theory predicted by his new quantum theoretical account, especially those predicted to occur at very low temperatures and pressures. Sackur was aided in these efforts by Haber and Friedrich Kerschbaum, who developed an idea of the American physical chemist Irving Langmuir into a new method for measuring very low pressures using the oscillations of a quartz fiber. The resulting quartz fiber manometer would resurface at the Institute after the war as a key part of innovative apparatus for research on low pressure gas reactions and luminescence.[42]

---

42 Stoltzenberg, *Haber*, p. 217.

Haber's Institute during the First World War

Haber took part in the widespread enthusiasm that accompanied German mobilization and entry into the First World War in the summer of 1914 and registered for voluntary military service at the beginning of August. He gave voice to his euphoria in a letter to Svante Arrhenius in Stockholm, writing:

> This is a war in which our entire people (*Volk*) is taking part with full sympathy and its utmost abilities. Those who don't bear arms work for the war, and everyone scrambles forward voluntarily for the slightest accomplishment. You know Germany too well not to know that such a unanimous commitment to a cause is only possible amongst us when all are conscious that the good of the nation must be defended through a just struggle. You should give no credence to the absurd fiction, according to which we are conducting a war out of military interests...but we now see it as our ethical duty, to take down our enemies with the use of all our strength and bring them to a peace that will make the return of such a war impossible for generations and give a solid foundation for the peaceful development of western Europe.[43]

When Harnack, as president of the Kaiser Wilhelm Society, called together all of the institute directors on August 12 to discuss possible consequences of the war for the work of the Society, Haber was already occupied by military concerns and sent Gerhard Just in his stead. Haber worked first as a scientific consultant in the Ministry of War for the Artillery Command and the Production Department, where his expertise in applied chemistry and ammonia synthesis were particularly valued. Representatives of German politics and industry had quickly realized that limited raw materials made a long war untenable for Germany. Nitrogen was of particular concern, as Germany relied upon Chilean saltpeter to supply both its fertilizer needs and its production of explosives and propellants, but the English sea blockade threatened to cut off this source. The war would also lead to shortages of myriad other raw materials and related bottlenecks in industrial production. Hence, German chemists faced the challenge of rationalizing use and production of these scarce materials or finding substitutes for them.

"In war, scientists belong to their Fatherland, like anyone, in peace, they belong to humanity."[44] Haber not only followed this maxim personally, he applied it to his entire institute and promptly redirected its resources toward projects relevant to the war. The conversion to military research projects proceeded surprisingly smoothly and without noticeable resistance. This raises the question whether the war euphoria alone eased the transition or whether something basic to the research policy of the Kaiser Wilhelm Society, especially its chemical institutes, enabled such a conversion. As Johnson has argued:

> True to his nature, Fischer stamped the scientific program of the Kaiser Wilhelm Society with a dual character. On the one hand, it was aimed at the most fundamental problems of natural science; but on the other, it was intended to produce solutions

---

43 Zott, *Haber*, p. 77.
44 Haber, *Industrie*, p. 252.

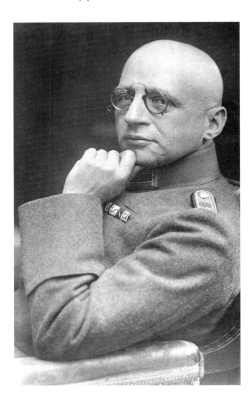

Fig. 1.14. *Fritz Haber in his captain's uniform, 1916.*

to technological problems of the highest national interest, particularly with regard to providing domestically available synthetic or artificial substitutes for imported materials.[45]

The development of the catalytic process for ammonia synthesis was already one realization of the desire to manufacture domestic substitutes for economically important imported goods. It is also a common belief among historians of the First World War that without the Haber-Bosch process the German military would have run out of munitions in 1915. Similar intentions led Haber to his wartime partnership with the Raw Materials Department of the War Ministry under Walther Rathenau, which eventually led him to research on chemical means for waging war. As Johnson pointedly summed up the progression: "the logic of Ersatz led to the problems of munitions, and eventually to poison gas."[46]

In the first months of the war, the Institute searched for ways to economize or provide substitutes for so-called "war materials" – substances required for the operation of firearms, artillery and other war machines; examples include toluene, glycerin and saltpeter. Gerhard Just made rapid progress in this field, in collaboration with Otto Sackur. Together they were able to demonstrate, through careful

---

45 Johnson, *Chemists*, p. 133.
46 Ibid., p. 189.

freezing point and boiling point measurements, that a combination of xylene and certain water-soluble fractions of crude oil could replace toluene as an anti-freeze in engines. Their discovery meant a savings of roughly 400 tons of toluene per month that could then be used in the production of TNT and other explosives and munitions. In the autumn and winter of 1914, Haber and his colleagues also took part in the development of respiratory irritants and tear gases in connection with the already mentioned conservation efforts.

Parallel to these efforts, Haber's Berlin colleague Walther Nernst, under a contract from the Supreme Army Command (*Oberste Heeresleitung*, OHL), sought to develop shells that "contained solid, gaseous or liquid chemicals that would harm the enemy or make them unable to fight." The grenades developed by Nernst proved relatively ineffective though. In a field test held near Nouvelle-Chapelle on 27 October 1914, the explosive charges in the grenades proved inadequate to disperse their irritant powder. Somewhat later, Hans Tappen, a chemist working for the Army Ordnance Office, developed a grenade which employed a liquid irritant rather than a powder and dispersed its payload more effectively while using a smaller explosive charge. These Tappen-Shells, or T-Shells as they were more commonly known, were part of the German arsenal by January of 1915.

In connection with their efforts to develop new and more effective explosives and propellants, Haber, Just and Sackur attempted to replace the irritant in the T-shells with a substance that would act as both irritant and propellant. They settled on using cacodyl chloride, which Bunsen had first synthesized in 1837, but which chemists had scarcely researched since, because it was such a powerful irritant and toxin. On 17 December 1914, during an experiment intended to improve further the effectiveness of the cacodyl chloride, there was an explosion in the laboratory. Just lost his right hand. Sackur was fatally wounded. Haber had left the room shortly before the blast and remained physically unharmed. Nevertheless, he was unsettled by the death of his highly-talented colleague and steadfastly honored his memory. Years later, Haber would arrange a secretarial position for Sackur's daughter at the Institute.

After the accident, research on cacodyl chloride at the Institute halted, but the explosion also marked a turning point not wholly ascribable to its tragic consequence: the end of significant research on explosives at the Institute and the beginnings of poison gas research. Sometime in the first half of 1915, Haber redirected research at his institute toward the needs of gas warfare. Unfortunately, the available sources do not provide precise answers as to when or how this occurred. What is clear at least is that Haber began in January of 1915 to plan the first gas attack of the war, which would take place at Ypres on 22 April. The attack released 150 tons of chlorine gas that not only threw the Allied forces into a panic but reportedly poisoned 7,000 and killed 350. In spite of the "success" of the attack, however, it became clear that the predominant westerly winds made gas cloud attacks too unpredictable and unreliable a tactic upon which to base a new method of warfare. This led Haber to a renewed interest in poison gas grenades and shells, which were not so dependent upon rapidly changing meteorological conditions.

Fig. 1.15. *Fritz Haber (second from left) overseeing the preparation of gas shells, circa 1917.*

The men of the gas brigades, who had been long mocked as "pest exterminators in uniform," and amongst whom numbered many Dahlem and Berlin scientists, including James Franck, Otto Hahn and Gustav Hertz, experienced an immediate improvement in status. Haber was even summoned before the Kaiser, who promoted him from staff sergeant to captain, a powerful recognition of the value of his efforts. This advance in rank appears to have further motivated Haber to the self-assigned task of promoting chemical warfare, first as chemical advisor to the Ministry of War, then beginning in November 1915, as head of the "Central Office for Chemical Concerns" in the Artillery Division. He essentially abandoned scholarly research and concentrated upon the problems of chemical warfare. In the words of his biographer Dietrich Stoltzenberg

> everything else in his life [faded] into the background. Wife and family now had almost no influence on his life. In fact, for him, family, friends, and acquaintances were just further sources of aid for his cause.[47]

But Haber paid a high price for his preoccupation with the war, his already strained relationship with his wife Clara Immerwahr, one of Germany's first female PhDs in chemistry, broke down completely, and on 2 May 1915 she shot herself with her husband's service weapon. The manner of Clara Immerwahr's death and its near simultaneity with the first gas attack at Ypres have only recently come to be seen as a protest against the new form of warfare ushered in by her husband

---

47 Stoltzenberg, *Haber*, p. 256.

Table 1.1. *Departmental structure of the Institute at the close of World War I, according to the report by Harold Hartley, 1921.*

Department	Capacity	Head
A	Development, Specification and Inspection of Respirator Face-pieces	Prof. Herzog
B	Technical Development and Testing of Offensive Appliances	Prof. Kerschbaum
C	Development of Respirator Drums and other Appliances	Prof. Pick
D	Synthesis of New Gases	Prof. Wieland
E	Pharmacological and Pathological Section	Prof. Flury
F	Inspection and Issue of Respirator Drums	Prof. Freundlich
G	Supply of Shell bodies, Fuzes etc.	Dr. v. Tappen
H	Trench Mortars	Dr. v. Poppenberg
J	Inspection and Issue of Toxic Gases	Prof. Friedländer
K	Particulate Clouds	Prof. Regener

The ongoing development of gas masks and filters, in increasingly close cooperation with industry, took place in Departments A (Herzog) and C (Pick). The work of these departments relied upon a steady exchange of knowledge between laboratory researchers and battlefield informants. Later in the war, prototypes from these departments would be tested against new toxins from Department B (Kerschbaum).[57] Members of Kerschbaum's department strove to find and identify substances with potential for use in gas cloud and shell attacks. Their work consisted of a mixture of literature research to identify substances with an optimal combination of noxiousness, low boiling point and high vapor density, and experiments on animals and volunteers from the Institute staff to confirm the irritating or toxic effects of these substances.[58] Department D (Wieland) focused specifically on deleterious arsenic and sulfur compounds, e.g. mustard gas, and performed primarily laboratory research, including attempts to synthesize new substances with effects analogous to known toxins and irritants. Research on the physiological effects of various poisons, including careful study of their relative toxicity, occurred in Department E (Flury) and relied upon extensive animal experimentation. It was also Flury and his collaborators who promoted use of the so-called "Haber Constant," the product of the concentration and the exposure

---

57 Hartley, *Report*, p. 39–42.
58 Ibid., p. 45.

time required to cause death. This constant aided early efforts to define limits on hazardous substances in the civilian sphere. Department J (Friedländer) was responsible for testing the quality of chemical weapons produced by industry, for which it employed predominantly classical analytical methods rather than measurements of physical constants.[59] Only in Department K, under Erich Regener, did techniques from physical chemistry play a central role. Regener's group used ultramicroscopes to study the small particles that constitute powders and smokes and their ability to penetrate existing gas mask filters.

Post-war assessments of the scientific value of this research by Allied representatives and later historians have been almost universally negative. In his own remarks on the subject during the 1920s, Haber emphasized the effectiveness and what he and others, also on the Entente side, called "humaneness" of chemical weapons, but nonetheless, explained to Allied agents that all of the important toxins used in the war had already been synthesized and studied before 1914 and that "no systematic progress had been made" in toxin research. In the same vein, Richard Willstätter reported to the Allies that he did not consider the synthesis research pursued at Haber's institute particularly serious.

Still, Haber was without question the driving force behind the centrally-directed development of chemical warfare in Germany, whose use during the First World War violated international law and elicited immediate and enduring moral criticism, and has thereby inadvertently come to personify the moral flexibility of scientific research. His efforts during the war argue for the impossibility of removing science from the socio-political context in which it is embedded and illustrate how quickly the fine line between these seemingly different realms can be crossed, sometimes with fatal consequences. Another keen example of this danger would come hardly twenty years later, this time with Haber as the victim rather than the perpetrator, and would similarly leave its distinct mark on the Institute.

---

59 Ibid., p. 50–52.

2 The "Golden Years" of Haber's Institute

According to the recollections of Otto Hahn, Haber had already begun to question the likelihood of a German victory in February of 1918.[1] He was worried both by American entry into World War I and by the growing gaps in the German budget, about which his membership on the Nitrogen Board and his military contacts kept him well informed. Still, Haber did not relent in his efforts to support the German military, and weapons production activities at the Institute continued up to and probably beyond the November Armistice. In a bid to stave off unemployment and economic depression, Josef Koeth, head of the newly established Demobilization Office, honored already established orders for munitions and kept large sectors of the armaments industry operating until the end of January 1919.

However, there was no doubt after November 1918 that the Institute would need to demobilize. This must have been a trying task given the number of employees, amount of equipment and number of buildings involved, but few documents remain relating to how it was achieved. In the first stages of the reorganization, Haber had a relatively free hand, in spite of demands from both the military and the civil authorities, as he was appointed head of the Chemistry Division of the Demobilization Office, which was responsible for the dismantling of chemical weapons production facilities. His chief aid at this post was Friedrich Epstein, who had also assisted Haber in the chemistry office of the Ministry of War and had been a guest at Haber's institute from its inception.[2] Later recollections also suggest that the Institute, and the community of Dahlem as a whole, remained relatively untouched by the widespread unrest of the autumn and winter of 1918–1919. Charlotte, Fritz Haber's second wife, even recalled a military regiment being stationed in Dahlem to help protect the Institute.[3] Possibly as a result, the Institute suffered relatively few material losses, a much better fate than it would face in the aftermath of the Second World War.

In February of 1919, free from his duties in the demobilization office, Haber turned wholeheartedly to the task of rebuilding his Institute and wrote to Carl Duisberg of BASF that

> I have made my institute into a kind of council of academics, in which all the professors with [the last initial] F have received membership.[4]

---

1 Hahn, *Leben*, p. 127.
2 Szöllösi-Janze, *Haber*, p. 411–413.
3 Haber, *Leben*, p. 138.
4 MPGA Abt. Va, Rep. 5, Nr. 860.

Fig. 2.1. *Transport of Institute members to an operation led by the Technical Emergency Help, end of 1918.*

This "council" included Paul Friedländer, Herbert Freundlich, James Franck and Ferdinand Flury. Friedländer, an organic chemist and expert on indigo derivatives, had directed the testing of industrially produced toxins during the war. He would lead a new department for pharmacological preparations. Freundlich, who had directed the department responsible for improving gas mask filters, would lead a department of colloid chemistry. Flury, a trained pharmacologist who was in charge of toxicity tests and animal experimentation during the war, would be joined by the entomologist Albrecht Hase, who had worked on pest control for the military administration during the war, in leading a department for zoology and pharmacology. Franck had not been in charge of a research department during the war, but worked with Haber as a member of the engineering corps unit handling gas warfare (pioneer regiments 35/36) until an illness sent him back to Berlin, where he helped with gas and gas mask testing. He was chosen to lead a new physics department at the Institute. Reginald Oliver Herzog, who had headed the department for improving gas mask materials, had already been appointed head of a department for textile research, the result of ongoing negotiations between the KWG and the textiles industry, in which Haber took an active role, aimed at eventually establishing an independent Kaiser Wilhelm Institute for Fiber Chemistry.[5]

---

5 Löser, *Gründungsgeschichte*.

buildings and grounds.[13] Scientific personnel would fluctuate rapidly in the following years but rarely dipped below two dozen active researchers and reached highs of over forty, although at no point after 1921 did more than twelve of these, including the department heads, receive official salaries from the Institute. The remaining researchers were privatdozents at one of the two Berlin universities, fellowship recipients, guests of the Institute from other academic establishments or under informal contracts with Haber.

Doctoral students also swelled the ranks of the Institute. Some doctoral candidates had assisted with weapons research during the war, but only after the war were doctoral candidates welcome to pursue research toward their degrees at the Institute. In part, the absence of doctoral students before the war had been due to concerns about admitting research institutes partly funded by industry into the existing educational structure, but Kaiser Wilhelm II had also stated explicitly in his speech at the centennial of the Berlin University that institutes of the Kaiser Wilhelm Society should not be influenced by teaching concerns, and something of this viewpoint resurfaced in 1926, when Walther Nernst proposed, unsuccessfully, that young researchers not be allowed to pursue their habilitation research at Kaiser Wilhelm Institutes.[14] In this vein as well, the buildings of the Institute for Physical Chemistry and Electrochemistry had not been designed to accommodate students, but rather with the expectation that researchers at the Institute would have at least completed their doctorate before arrival. In light of the financial hardships of the early 1920s, however, Haber suggested to Adolf von Harnack that, not only at his own institute but also at the neighboring institutes for chemistry and biology:

> in the case of scientific personnel ... the only way to achieve significant savings, is to work to a greater extent with young people, for whom the work is an integral part of their training, that is with doctoral students.[15]

Initially, some of the doctoral students at the Institute were under the supervision of Haber, while others performed research at the Institute but were supervised by one of the professors at Berlin University or the Technical University (*Technische Hochschule*, TH) in Charlottenburg, often Max Bodenstein or Arthur Rosenheim. Once Herbert Freundlich became an honorary professor at Berlin University (1925), and later also at the Technical University (1930), he too oversaw doctoral students, including Vera Birstein (Ph.D. 1926). When Michael Polanyi joined the Institute, he also advised doctoral students, notably fellow Hungarian Eugene Wigner (Ph.D. 1925). Classroom teaching established another new bridge between the Institute and the Berlin universities. Haber was obligated to teach a one-semester course in chemical technology in connection with his professorship at Berlin University, but his absence from Berlin during the early 1920s (see below) meant that he fulfilled this obligation at best sporadically, leading to some tension with the

---

13 Haber, Jahresberichte 1921, MPGA Abt. Va, Rep. 5, Nr. 1912.
14 Haber to Schmidt-Ott, 21 June 1926, MPGA Abt. Va, Rep. 5, Nr. 1687.
15 Haber and Harnack, 19 January 1920, MPGA Abt. Va, Rep. 5, Nr. 1905.

Fig. 2.5.   *Institute colleagues on their way to coffee by the Schlachtensee. Left to right: Friedrich Epstein, Paul Goldfinger, Ladislaus Farkas and Hartmut Kallmann (driving), early 1930s.*

University Administration. In the course catalog one finds listed only "chemical and physical experiments, e.g. scientific work at the Kaiser Wilhelm Institute for Physical Chemistry." However, Freundlich lectured on sundry topics in colloid chemistry, and Michel Polanyi led a course on metals and their processing. Karl Friedrich Bonhoeffer taught on various topics in physical chemistry and atomic physics between winter of 1927 and early 1930, while Paul Harteck offered courses on photochemistry and inorganic chemistry in 1931/32 and 1932/33. Although, early in 1932, Harteck informed his mentor Bonhoeffer of his fears that, "[my] lecture will perhaps go under. I have only 6 attending."[16] Nevertheless, this marked a much closer integration of the Institute into the existing academic system and a concession of sorts to Humboldt's principle of the unity of teaching and research over the Kaiser's call for independence from the influence of educational goals.

Two further lasting changes at the Institute that date to the first days after the war were the inauguration of the so-called "Haber colloquia" and the promotion of affiliates of the Institute to scientific membership. Archival material regarding the early colloquia is sparse. The organizers did not leave behind a unified list of presenters and participants. However, the colloquia feature prominently in the recollections of several famous scientists and ranked with the Wednesday Physics Colloquia at Berlin University and the Friday Colloquia of the Physical Society as one of the key gathering points for physical scientists in Berlin. The biochemist David Nachmansohn remembered them as "one of the most striking and valuable

---

16 Harteck to Bonhoeffer, 4 May 1932, PHP, 1 : 17.

Physical Chemistry in Berlin

During the decades surrounding the turn of the 20th Century, Berlin was a leading worldwide center of scientific research. Historians of science speak in particular of the "Great Berlin Physics," which, in the period 1870–1930, was marked by the presence of such distinguished physicists as Hermann Helmholtz, Max Planck, Albert Einstein and Erwin Schrödinger. A similar epithet is well in order for Berlin physical chemistry. In the words of Paul Harteck, "in the period from 1919 to 1933, many of the outstanding physical chemists of Germany were concentrated in Berlin. These included Nernst, Haber, Bodenstein, Volmer, and Bonhoeffer ... a center of physical chemistry developed which was unequalled anywhere in the world at the time."

Walther Nernst, Albert Einstein, Max Planck, Robert A. Millikan, Max von Laue, Berlin 1931.

The burgeoning of physical chemistry in Berlin started in 1895 with a call to Jacobus Henricus van't Hoff to assume a research professorship at the Prussian Academy. Van't Hoff, with his fundamental contributions to chemical kinetics, chemical equilibria and affinity, was one of the founders of physical chemistry and the first Berlin Nobel Laureate (1901). Then in the Spring of 1905, Walther Nernst left the Institute for Physical Chemistry established for him in Göttingen and accepted a professorship at the Berlin University. The year after Nernst arrived in Berlin, Wilhelm Ostwald, one of the central figures in physical chemistry, retired from his post in Leipzig, thereby clearing the way for Berlin's dominance of the field. Moreover, Nernst heralded his arrival in Berlin with a roar by enunciating the Third Law of Thermodynamics the year of his arrival. The experimental and theoretical basis for the Third Law remained at the focus of his research in subsequent years, which also cemented Berlin's position as one of the early centers of the young quantum theory.

When building up the Institute in Dahlem, Haber made limited use of the advantages offered by the Berlin physical chemistry community. Van't Hoff died the year Haber arrived in Berlin, while Haber and Nernst had previously engaged in a heated dispute concerning the thermodynamics of ammonia synthesis and were not close personally. Moreover, Haber brought most of the initial Institute staff with him from Karlsruhe.

Physical chemistry research in Berlin reached its peak following the appointments of Max Bodenstein and Max Volmer. In 1922, the TH Charlottenburg appointed Max Volmer head of an Institute for Physical Chemistry and Electrochemistry. During the war, in collaboration with Otto Stern, Volmer had worked on the kinetics of inter-molecular deactivation processes, e.g. the quenching of fluorescence, governed by the Stern-Volmer relationship. In the Weimar era, Volmer's interest shifted toward crystal growth and the kinetics of phase transitions. Bodenstein returned to the Berlin University in 1923, where he had served before as *extraordinarius* professor from 1906 to 1908, to replace Nernst. Nernst left the University in 1922 but never left the city. He served two years as President of the Imperial Institute for Physics and Technology in Charlottenburg then returned to the University to head its Physics Institute as successor to Heinrich Rubens. Bodenstein had introduced the concept of a chain reaction consisting of elementary reaction steps while studying the kinetics of the light-induced reaction of hydrogen and chlorine gases with his assistant Walter Dux in Hannover, and he continued this line of research after his return to Berlin. Bodenstein's earlier investigations of the photochemistry and kinetics of gas reactions laid much of the groundwork for the research on these topics pursued by Haber and Polanyi. Bodenstein and Haber were also in personal contact through their professorships at the Berlin University, and they jointly fostered a symbiotic relationship between their two institutions during the 1920s. While the Dahlem Institute greatly profited from a steady stream of first-rate doctoral students, the curriculum of Berlin's universities was enriched by courses offered by both junior and senior members of the Institute.

experiences in his scientific formation."[17] The colloquia were known for their relatively informal atmosphere, the very direct questions of their audience members and the disciplinary and national diversity of their speakers. The topics of the colloquia ranged "from the helium atom to the flea," and amongst the many guest speakers at the colloquia numbered physicist Peter Debye, chemist Richard Willstätter, biologists Otto Warburg and E. Newton Harvey, and biophysicist Selig Hecht. Niels Bohr offered a special "boss-free" colloquium during his visit to Berlin in April of 1920, at which participants discussed the latest developments in atomic theory. Only those below the rank of professor were invited, in order to free the discussion as much as possible from considerations of academic standing; however, all of those pictured below rose quickly to the rank of professor and five of them received Nobel prizes. Initially, Herbert Freundlich took primary responsibility for organizing the colloquia, which took place at the Institute every second Monday. In 1929 the colloquia moved to the newly opened Harnack House, a few blocks

---

17 Nachmansohn, *Pioneers*, p. 184.

Fig. 2.6.   *"Boss-free" colloquium held during Niels Bohr's visit to Berlin, April 1920. Left to Right: Otto Stern, Wilhelm Lenz, James Franck, Rudolf Ladenburg, Paul Knipping, Niels Bohr, Ernst Wagner, Otto von Baeyer, Otto Hahn, George von Hevesy, Lise Meitner, Wilhelm Westphal, Hans Geiger, Gustav Hertz, Peter Pringsheim.*

from the Institute, and in 1930 Michael Polanyi took charge of organizing the colloquia, but their intellectual character appears to have remained unchanged.

As to the promotion of affiliates to Scientific Members of the Institute and by extension to membership in the Kaiser Wilhelm Society, this was an option available to the Scientific Director and the Board of the Institute since its founding, but which they only chose to begin exercising after the war. In 1918 Haber proposed that his long-time assistants in administering the Institute, Friedrich Epstein and Friedrich Kerschbaum, be admitted as Honorary Members of the Institute on the basis of "service in times of war and peace." The following year, Reginald Oliver Herzog, head of the Department for Fiber Chemistry received membership in the Society, as did Herbert Freundlich, head of the Colloid Chemistry Department. Michael Polanyi had already become Scientific Member of the Institute for Fiber Chemistry before he took over as a department head at Haber's institute in 1923, and Rudolf Ladenburg, who took over a new Physics Department late in 1924, also received scientific membership in the Institute; although there was no clear rule, as there would be in later years, relating scientific membership to the direction of a research department. When the Kaiser Wilhelm Society made provisions for external scientific members in 1926, former Institute coworkers Ferdinand Flury, James Franck, Johannes Jaenicke and Gerhard Just became external Scientific Members

Fig. 2.7. *Institute Library, circa 1930.*

of the Institute, and Kerschbaum and Epstein had their honorary memberships converted to external memberships.

As this uptick in memberships shows, the Institute was quick to recover after the collapse of Haber's academic council. Haber never fully accepted the demise of physics at the Institute. When James Franck finally left Berlin, in March of 1921, his colleagues, Paul Knipping and Walter Grotrian, both working toward their habilitations, remained at the Institute, as did the doctoral candidates Franck had overseen, amongst them Hertha Sponer, who would become Franck's second wife some twenty-five years later. Fritz Reiche, then a privatdozent at Berlin University, also stayed on as a "theoretical advisor" at the Institute. Reiche was a gifted theorist sometimes called "the little oracle," in a play on Niels Bohr's nickname, and wrote one of the first textbooks on quantum theory.[18] He was also a close friend of Rudolf Ladenburg, who visited him frequently at the Institute before later taking over its Physics Department. In addition, a plan developed to promote Paul Knipping, co-discoverer of the diffraction of X-rays by crystals, to head a new Physics Department, but this plan collapsed when Knipping appeared unable

18   Bederson, *Reiche*. Wehefritz, *Reiche*. Schaeffer, *Reiche*.

Fig. 2.8. *"Department M," 1924. Left to Right: Schmitzspahn (lab-asst.), F. Epstein, Groth (secretary), H. Eisner, Wolff, H. Lehrecke, F. Haber, Matthias, Ehlermann, W. Zisch, Bahr (lab-asst.), J. Jaenicke, Kuckels (lab-asst.).*

to complete his habilitation and his relations with Haber turned acrimonious.[19] Haber was not able to secure a new head for physics research at the Institute until September of 1924, when Ladenburg accepted a permanent post; however, Haber viewed this as a key appointment, in that "atomic structure" remained one of the main foci of "pure scientific" research at the Institute, in harmony with the opinions Haber expressed before the war regarding the potential import of quantum theory to chemistry.[20]

As for what Haber termed "commercial" research at the Institute, Freundlich began a collaboration with Alexander Nathansohn before the collapse of the initial plan for the Institute, in which they investigated the use of solutions and colloids in metal refining. Their research provided the Institute with industrial support in excess of the direct needs of the project and turned out at least one viable patent whose proceeds benefited the Institute.[21] Meanwhile, Haber undertook an ambitious and secretive project to develop a means for separating gold from seawater on an industrial scale.[22] His end goal involved once more putting his scientific knowledge at the service of the German state, which faced reparations from the First World War amounting to approximately 132 billion gold marks or 50,000 tons of

---

19 Glum to Harnack, 6 April 1923, MPGA Abt. Va, Rep. 5, Nr. 1916.
20 Haber to Harnack, 9 June 1923, MPGA Abt. Va, Rep. 5, Nr. 1916; Haber, *Zeitalter*.
21 Haber to Koppel, 1 December 1921, MPGA Abt. Va, Rep. 5, Nr. 1703.
22 Cf. Hahn, *Gold*.

gold. Though the development of such an economical separation process appears farfetched to modern experts, this is in large part a result of Haber's own efforts. Multiple estimates of the gold content of seawater current at the time, including one by Nobel Laureate Svante Arrhenius, suggested that separation was feasible, and William Ramsay, another Nobel Laureate in chemistry, had already covertly begun research into recovering gold from seawater. For its size, the gold from seawater project consumed relatively little of the Institute's resources. Doctoral students carried out many of the laboratory investigations, including the preliminary research, and in November of 1922 Haber entered into a contract with two industrial sponsors, Degussa (*Deutsche Gold- und Silber-Scheideanstalt vorm. Roessler)* and the *Metallurgische Gesellschaft (Lurgi) Frankfurt am Main*, who were willing to cover essentially all expenses of the research in return for shares in any resulting patents. This marked the establishment of a new "Department M" at the Institute, which grew to include approximately 20 members and was led by Johannes Jaenicke, a chemist who had been at the Institute since roughly half-way through the war. Members of the new department undertook four separate research voyages, only to discover that the concentration of gold in seawater (averaging roughly 10 ppt) was not high enough to allow economical recovery. Proving this involved the development of innovative analytic chemistry procedures, but no patents were forthcoming, and the department dissolved in 1926 with Jaenicke leaving the Institute to take a position at the Metallgesellschaft.

Rebuilding within the Kaiser Wilhelm Society

Neither of these industrial projects was able to cover the growing shortfalls in the Institute budget. From spring of 1919 to spring of 1920, Institute expenditures amounted to approximately 599,000 marks, already 146,000 marks more than its income. The following year expenses rose to 877,000 marks, while the total income of the Institute actually sunk to 386,000, a total shortfall of 481,000 marks. The vast majority of the increase in the budget came from attempts to compensate for inflation by increasing salaries at the Institute. The following year was only worse: personnel expenses nearly doubled, from 530,600 marks to 992,400 marks,[23] and the 10 year commitment by the Koppel Foundation to provide a 35,000 mark annuity for the Institute came to an end, although by that point inflation had already reduced the grant from covering roughly 50 % of the Institute's expenses in 1911–1912 to roughly 5 % of its expenses in 1921–1922. The hyperinflation of 1922–1923 was, of course, cataclysmic for an institute on a fixed annual budget, and already in January 1923 Haber was forced to ask the General Secretary of the Kaiser Wilhelm Society, Friedrich Glum, for further funds to cover the fiscal year ending March 31, as the budget had risen to 5,950,000 marks, and Haber

---

23 Haber to Glum, 2 March 1922 with "Aktenvermerk," MPGA Abt. Va, Rep. 5, Nr. 1792.

complained that even if this amount were transferred immediately, the resulting sum would only cover one-quarter of the budget by the time it arrived.[24]

Other Kaiser Wilhelm Institutes faced similar difficulties, and discussions of government intervention to balance the budgets of the Society and its member-institutes began in earnest in the summer of 1920. From the outset support for Haber's institute was part of the plans under discussion, although the Institute remained financially independent of the Kaiser Wilhelm Society. As a temporary measure the Institute received 500,000 marks from federal funds designated for the Emergency Association of German Science (*Notgemeinschaft der deutschen Wissenschaft*, NG) to cover its deficit from fiscal year 1920/21, but when a more permanent agreement was reached in 1922, whereby the Federal and Prussian Governments shared responsibility for covering the deficits of the Society and its member-institutes, the Institute for Physical Chemistry and Electrochemistry was not included.[25] Haber discussed with Max Donnevert of the Federal Ministry of the Interior possible solutions to this oversight and Donnevert proposed three options: absorbing the Institute into the Kaiser Wilhelm Society, bringing the Institute directly under the auspices of the Federal Government, or obligating the Emergency Association to continue covering the Institute's deficit. Haber was reluctant to sacrifice the independence of his institute, but in the end, limits on the Federal budget, possibly the result of spending restrictions by the Allied powers, made the entrance of the Institute into the Kaiser Wilhelm Society the only politically viable solution, and in December 1922 the Prussian and Federal Governments contributed equally to covering 6 million marks of the Institute's deficit per the standard agreement regarding support for member-institutes of the Society. The official date of the entrance of the Institute into the Kaiser Wilhelm Society is unclear though, as a new curatorial board for the Institute, in keeping with Society statutes, was not formed until May of 1923. This marked the definitive end of the arrangement in which the Institute conformed to the ideals of the Kaiser Wilhelm Society but remained, in fact, independently funded and administered. Moreover, as responsibility for support of the Kaiser Wilhelm Society shifted from the Prussian to the Federal Government over the next decade, reaching roughly 80% federal funding by 1932, the previously symbolic national character of the Institute took on a concrete, fiscal aspect.

As the Institute moved under the auspices of the Kaiser Wilhelm Society it was in a period of rapid internal transition. In fiscal year 1922/23, the Institute employed two department heads, Haber and Freundlich, as well as seven scientific assistants, not counting Department M under Jaenicke, whose expenses were paid by its industrial sponsors. In addition, Margarethe von Wrangell ran an independent research group that was hosted by the Institute but supported by funds from the Kaiser Wilhelm Society and from the "Japan Committee" of the

---

24 Freundlich, "Übersicht über Etatsjahre 1914, 1919, 1923," MPGA Abt. I, Rep. 1a, Nr. 1179.
25 Cf. Szöllösi-Janze, *Haber*, p. 500 ff.

Haber and Science Policy

While prior to WWI, Fritz Haber was primarily a beneficiary of science policy initiatives and organizational efforts by other high-ranking scientists and officials, in the Weimar Republic he put his own reputation as a well-established scientist, Nobel Laureate and respected participant in the Great War to use in bolstering public support for scientific research in Germany. Science funding was particularly scarce after the Great War, since the all but empty public coffers left after the collapse of the German Empire and the concomitant political and economic turbulence were insufficient to even maintain the existing research establishment. Voices all over the country spoke of the plight of German science and Haber joined in the chorus, e.g., in an article for Berliner Tagblatt from 7 March 1920 entitled "The Crisis of German Science."

Seated camelback, from left to right, Setsuro Tamaru, Charlotte and Fritz Haber visiting the pyramids of Giza during Haber's world tour in 1924.

In parallel, behind the scenes, Haber courted the tycoons of the chemical industry and convinced some of them to bestow their lasting support for university research. He also contributed to the establishment of the Society for the Funding of Chemical Education and the Emil Fischer Society for the Funding of Chemical Research. However, Haber perceived all this as inadequate and, above all, too specialized and sought a more comprehensive solution. This he found, together with Friedrich Schmidt-Ott and others, in the Emergency Association for German Science (*Notgemeinschaft der Deutschen Wissenschaft*), today's German Science Foundation (*Deutsche Forschungsgemeinschaft*), which was established on 30 October 1920. While Haber worked mainly behind the scenes and influenced the structure and organizational principles of the planned Emergency Association, Schmidt-Ott did the public relations work and chaired the corresponding founding committees. It is to Haber's credit that the Emergency Association broke with the traditions of Wilhelmine Germany and was not to be administered by the state bureaucracy. Instead, the Emergency Association was run from the outset by the scientific community itself, through representative panels. On Haber's initiative, the members of these panels, who decided upon the funding of specific projects, were elected rather than appointed. But Haber's further attempts to limit, through democratic structures, the autocratic powers of the president of the Emergency Association remained unsuccessful. The democratization of the hierarchical structures of the German academic establishment was a broader aim for Haber. He had also made an unsuccessful proposal that the members of the standing committees of the Bunsen Society be elected and that they be limited to just two terms of service. However, a 1928 petition to the KWG that he initiated, signed by 13 out of 30 KWG directors, did lead to the establishment of the Scientific Council (*WissenschaftlicherRat*) of the KWG, a body which gave a voice to the scientific members of the institutes, and remains one of the key organs of today's MPG.

Haber's commitment to science policy also had a strong international component. This is particularly evident in the relations between Japan, the KWG and Haber himself. In 1913 Japanese chemist and industrialist Jokichi Takamine pointed out the need for a Japanese national research institute. The following year, the industrialist Eiichi Shibusawa, consciously following the example of the KWG, petitioned the Japanese government for the establishment of an "Institute for Chemical Research." The war delayed the realization of this proposal, but with time, physics was added to the purview of the planned institute and on 20 March 1917 the *Rikagaku Kenkyusho* (RIKEN) was officially founded. The first edifice of RIKEN, No. 1 Building, was modeled after the KWI for Physical Chemistry and Electrochemistry. Setsuro Tamaru, who had returned to Japan after working with Haber in Karlsruhe and Dahlem and then with Theodore W. Richards at Harvard, played a major role in its construction, and when No. 1 Building opened in 1921, after four years of construction delays, it included, among other apparatus, facilities for the kind of high and low pressure gas research that contributed so much to the fame of Haber and his institute.

Haber's ties to Japan and his involvement with the Emergency Association came together in 1922 through the person of Hajimé Hoshi, founder of Hoshi Pharmaceuticals and Hoshi University and friend to several prominent Japanese political figures including Count Shimpei Goto and German Minister Wilhelm Solf. In fiscal year 1921/22 Hoshi pledged 80,000 yen, roughly 160,000 marks, in support of the Emergency Association. This money was immediately channeled into existing projects. However, in 1922, Hoshi visited Berlin, met Fritz Haber and made two further donations to the Emergency Association of 10,000 and 2,000 yen each. These were combined to form the Japan Committee of the Emergency Association, whose board Haber chaired. During his visit Hoshi also invited Haber to come to Japan. Haber accepted and made Japan part of a world tour he took in the fall and winter of 1924–1925. In Japan, Haber developed ties with Count Goto and with Minister Solf, with whom he collaborated in establishing the Japan Institute in Berlin, one of two parallel institutes (the other in Tokyo), intended to promote cultural exchanges between the two countries. Haber also served as first Chairman of the Board of Trustees of the Japan Institute.

Fig. 2.9. *Margarethe von Wrangell (1877–1932), 1921.*

Emergency Association. Von Wrangell arrived at the Institute in 1922 to pursue research on the use of phosphates in fertilizers, attracted in part by the acclaim of Haber amongst agricultural chemists. When plans for a longer-term post for her within the Kaiser Wilhelm Society broke down in 1923, she returned to Hohenheim Agricultural University. There she became chair of an institute for plant nutrition, making her the first female full professor in Germany.[26] In June of the same year, Haber requested approval from the Kaiser Wilhelm Society for the appointment of two new department heads.[27] Michael Polanyi, from the neighboring Institute for Fiber Chemistry, was to take charge of research in physical chemistry, a task for which Haber felt he no longer had sufficient time given his duties in Department M. Rudolf Ladenburg, then working at Breslau as a privatdozent and laboratory assistant, but formerly a guest of the Institute, was to direct a new Physics Department, as it was clear by that point that Paul Knipping would be leaving the Institute; Knipping would make an academic career for himself first at Heidelberg University then at Darmstadt Technical University. Polanyi joined the Institute on 1 September 1923, but Ladenburg did not officially take up his new post until a year later, on 15 September 1924.

After the appointments of Polanyi and Ladenburg, the shape of the scientific staff at the Institute was essentially stable. In addition to the four department heads, the Institute employed eight scientific assistants: four under Haber, two under Freundlich, one under Ladenburg and one designated for Polanyi beginning in fiscal year 1925/26. This personnel schema lasted up to the summer of 1932, when preparations began for Ladenburg to move permanently to the physics department at Princeton University, but the schema was primarily a budgeting tool. In most cases, the assistantships were assigned to researchers already affiliated with the Institute as needed. Through 1926/27 Freundlich was assisted by

---

26 Cf. Andronikow, *Wrangell*; Wrangell to Haber, 27 August 1923, MPGA Abt. Va, Rep 5, Nr. 1703.
27 Haber to the Kuratorium of the KWI, 11 June 1923, MPGA Abt. Va, Rep. 5, Nr. 1916.

Fig. 2.12. *Meeting of the "Hoshi Committee" of the Emergency Association. From left to right, seated: F. Haber, W. Schlenk, M. Planck, R. Schenck, R. Willstätter; standing: H.D. v. Schweinitz, K. Stuchtey, Mudra, M. Donnervert, O. Hahn, H. Krüss.*

Institutes, and by the autumn of 1925, possibly in response to a fire in the main building of the Institute in June, Haber had begun discussions with Friedrich Glum regarding necessary renovations at the Institute. Haber initially hoped to convince Leopold Koppel, who had pledged to support the Institute with 15,000 marks a year for 10 years, to instead offer the promised funds in a single grant for renovations.[35] This plan failed, and instead, in September of 1926, Haber proposed a list of renovations that he felt the Renovation Fund of the Kaiser Wilhelm Society could support. In addition to roof and flooring repairs and similar maintenance work, the plan included the closing of the "Kaiser's Entrance" facing Faradayweg, the expansion of the administrative offices, and the construction of two new buildings, a glassblowers workshop and a new X-ray laboratory. The Federal Government approved 50,000 marks in support of the project and the KWG chose Carlo Sattler of Munich, who also designed the nearby Harnack House, as the chief architect.[36] Haber chose Siemens AG to provide the electrical installations, just as they had for the original Institute. The project budget included some 95,000 marks for the new laboratory and 20,000 marks for the glassblowers' workshop, in addition

35 Haber to Glum, 6 October 1925, MPGA Abt. Va, Rep. 5, Nr. 1907.
36 Glum Vermerk, 26 March 1927, MPGA Abt. I, Rep. 1a, Nr. 1208.

Fig. 2.13. *Fire in the main building, 3 June 1925.*

to 40,000 marks worth of building repairs and minor alterations. However, the project went almost 45,000 marks over budget, in part because Haber added a 27,000-mark high-voltage apparatus to the plan at the last minute.[37] The opening of the new buildings was celebrated on 29 September 1928, in conjunction with the inauguration of the new Kaiser Wilhelm Institute for Breeding Research.[38] In connection with the expansion, the Institute added a glassblower and a librarian to its payrolls, but these expenses would be offset in part by the opening of Harnack House the following year, which allowed the Institute to dismiss the cook it had employed in its "club room."

According to Haber, however, the new laboratory was essentially full upon completion. In 1927, Karl Bonhoeffer, Hartmut Kallmann and Hans Zocher all received

---

37 Haber to Morsbach, 8 February 1928; Sattler to Morbach, 16 February 1928. MPGA Abt. I, Rep. 1a, Nr. 1209.
38 Protocol KWG Senat, 1 June 1928, MPGA Abt. Va, Rep. 5, Nr. 1928.

their habilitations, and Karl Weissenberg, formerly of the Institute for Fiber Chemistry, became a guest of the Institute. In Haber's opinion, as habilitated scholars, all of these men, along with Georg Ettisch of the Colloid Chemistry Department, qualified to direct their own "research groups," i.e. to pursue independent topics of research and direct their own assistants and graduate students.[39] These groups filled the Institute to capacity and the total scientific staff of the Institute remained stable between April 1928 and June 1929. Herbert Freundlich apparently disagreed with Haber somewhat concerning the capacity of the Institute and complained in 1929 that the two assistants and 25 other researchers active in his department at the time markedly exceeded its capacity.[40] Be that as it may, publications from the Institute kept pace with the personnel expansion, rising in number from 50 to 70 articles and growing in total pages from 523 to 834 between 1926 and 1928, an increase of roughly 50%, on a par with the increase accompanying the earlier establishment of the new departments under Polanyi and Ladenburg.

This final expansion occurred, however, at the same time as the new German Minister of the Interior Carl Severing and the Prussian Culture Minister Carl Becker were working on designing a new, coherent science policy that would reduce expenditures for research and give the government greater say in how these funds were spent. For the Institute, this meant that in June of 1929 Friedrich Glum wrote to warn Haber that in the following year, if not sooner, the Institute would face significant cutbacks in Kaiser Wilhelm Society support.[41] The situation was only made worse by the Great Depression, so that between fiscal years 1930/31 and 1931/32 the contributions from the Kaiser Wilhelm Society, which accounted for roughly 75% of the Institute income, were reduced from approximately 342,000 to 291,000 marks. The next year, support from the Society would be reduced another 12,000 marks.[42] One of the immediate results of these cuts was a reduction in the total researchers at the Institute from roughly 65 in January of 1931 to 42 in January of 1932.[43] Although many of the researchers who were asked to leave, or not replaced upon their departure, were not paid by the Institute, the budget could no longer provide for their equipment expenses.

In fact, equipment needs were a growing concern at the Institute. Although the buildings had been renovated, most of the apparatus at the Institute, outside of the new X-ray laboratory, had not been updated since the establishment of the Institute. According to Haber, by 1929 the "vacuum pump did not work," the high-pressure apparatus was being used as a teaching aid by the locksmith, and some of the working equipment was "only of historical interest."[44] Haber received approval from the Senate of the Kaiser Wilhelm Society for 475,000 marks of special funding to be paid over the course of five years that would have allowed him to

---

39 Haber to the KWG, 14 June 1929, MPGA Abt. Va, Rep. 5, Nr. 1908.
40 Lauder Jones, 14 November 1929, p. 11. RAC, RF 12.1, Nr. 64.
41 Morsbach Aktennotiz, 6 October 1929, MPGA Abt. I, Rep. 1a, 1180.
42 Jahresrechnungen für 1930, 1931, 1932, MPGA Abt. I, Rep. 1a, Nr. 1181.
43 Weaver Diary, 20 January 1932, RAC, RF, RG 1.1, 717D, Nr. 110.
44 Haber to the KWG, 14 June 1929, MPGA Abt. Va, Rep. 5, Nr. 1908.

Fig. 2.14. *Left to Right: Karl Friedrich Bonhoeffer, Ladislaus Farkas (seated), Paul Harteck, Hans Reichardt (?), circa 1928.*

update the apparatus at the Institute, but the financial crisis put an end to the plan, and the Institute instead received only a one-time payment of 72,000 marks.[45] In response, Haber and Freundlich turned to the Rockefeller Foundation for help. The Foundation had recently supported an expansion of Felix Klein's mathematical Institute at Göttingen and pledged some $ 655,000 in 1930 to construct laboratories for the Kaiser Wilhelm Institutes of Physics and of Cellular Physiology, but piecemeal equipment grants were not common practice for the Rockefeller Foundation.[46] Nevertheless, the Foundation approved two equipment grants for the Institute. The first, in 1928, provided approximately 10,250 marks to update apparatus in the Colloid Chemistry Department. The second, in 1932, provided roughly 30,000 marks, one-third of which was earmarked for a new liquid air generator, while the remaining two-thirds supported the construction of a novel form of ion accelerator, designed by Harmut Kallmann and capable of accelerating

---

45 Haber to Planck, 16 June 1932, MPGA Abt. I, Rep. 1a, Nr. 1181.
46 Cf. Macrakis, *Rockefeller*; Kohler, *Partners*.

Fig. 2.15. *Planting of the Haber Linden Tree on the occasion of Fritz Haber's 60th Birthday, 9 December 1928.*

lithium ions to energies of 1.65 MeV, a patent-worthy improvement upon existing accelerators.[47] Nevertheless, these two grants covered only a very small portion of what Haber saw as vital modernizations of Institute apparatus.

The same year it provided its final equipment grant to the Institute, the Rockefeller Foundation also provided fellowships for Paul Harteck to pursue post-doctoral studies in Cambridge with Ernest Rutherford and for Hans Kopfermann

---

47 Brix, Ingwersen, Jaeschke, Repnow, *Beschleuniger*, p. 266. Weiss, *Spannung*.

to do the same with Niels Bohr in Copenhagen. The outcomes of these grants highlight the abruptness of the changes that would occur at the Institute after the passage of the Law for the Restoration of Professional Civil Service, in April 1933, and the subsequent complete restructuring of the Institute. Herbert Freundlich and other departing members of the Institute in control of equipment bought with Rockefeller funds wanted to take it with them upon their departure, an arrangement discussed in earnest by representatives of the Rockefeller Foundation and tentatively approved by Max Planck, president of the Kaiser Wilhelm Society, but initially strenuously opposed and almost prevented by representatives of the National Socialist Party.[48] Paul Harteck and Hans Kopfermann, on the other hand, found themselves stuck abroad while the Institute to which they were supposed to return was utterly redefined. This led to extensive correspondence between Harteck, Bonhoeffer and Planck concerning the responsibilities of the Kaiser Wilhelm Society to displaced scholars, before the issue was made irrelevant for Harteck by an offer from Hamburg University. Kopfermann, similarly was lucky enough to receive an assistantship with Gustav Hertz at the Technical University in Charlottenburg.

Further details of the transition of the Kaiser Wilhelm Institute for Physical Chemistry and Electrochemistry to the service of the National Socialist regime and the attendant changes in its structure and its research are reserved for the next chapter. The previous examples are provided only to give an impression of the jarring nature of the transformation that began in April 1933. Given the ostensible familiarity of many aspects of the Weimar Era Institute, it is tempting to see this "Golden Era" of the Institute as something of a direct forerunner of the modern Fritz Haber Institute, whose "natural" development was hindered by the subsequent vicissitudes of Germany as a whole, and Berlin in particular. There is, however, no clear continuity between the "modern" aspects of the Weimar Era Institute and the present Institute, and the unfamiliar context in which familiar looking structures at the Institute developed sometimes lent them significantly different meanings. However much Haber's "academic council" might resemble a board of scientific directors, and however much the subsequent research departments and research groups might look like modern organs of a scientific research facility, their existence depended completely upon Haber's whim. To play upon a trope from German social history, this was "research diversification from above" not from below. Similarly the size and diversity of the Institute staff was the result of not *only* a dedication to meritocracy on Haber's part but also the constant financial hardships of the Institute and its unusual place in the German research community. Even stripped of such an illusion of modernity, however, the Weimar era structure of the Institute remains exceptional on the basis of the sheer number of distinguished researches and researchers it produced and the enthusiasm with which they were greeted by the scientific community.

---

48 Cf. Szöllösi-Janze, *Haber*, p. 669–673.

Research Orientation:
Colloid Chemistry and Atomic Structure

In broadest terms, research activities at the Institute during the Weimar era revolved around multiple, overlapping series of theory-savvy, experimental investigations. In testament to this mode of research, distinct lines of investigation at the Institute frequently shared experimental objects or apparatus and often explored similar models of atomic structure, molecular cohesion or reaction kinetics, but bear no marks of an overarching research strategy or single, clearly-defined method. Articles by Institute members clustered around a limited array of topics: phenomena such as luminescence and capillarity, substances such as iron oxide colloids and mercury vapors, and techniques such as low-pressure manometry. However, the connections both within and between these research clusters were diverse and diachronic, resulting from personal connections and reliance upon shared resources. As a consequence, publications from the Institute during this era evince more a 'family resemblance' than a unitary logic of organization.

Both of the fields of "pure" research that Fritz Haber singled out as the strong points of his institute in 1923, colloid chemistry and atomic structure research, were broad and multifaceted, and neither fits neatly with our present

Fig. 2.16. *Fritz Haber amongst colleagues, from left to right, standing: Paul Goldfinger, unknown; seated: Graf von Schweinitz, Ladislaus Farkas, circa 1930.*

understanding of physical chemistry.[49] Haber and Freundlich were hardly alone, however, as vocal advocates of the fundamental significance of these fields during the 1920s. Both fields appeared to offer promising foundations for "general chemistry," one of the epithets the father of physical chemistry Wilhelm Ostwald preferred for describing the fledgling discipline. Moreover research in both fields, at least as it was pursued at the Institute for Physical Chemistry and Electrochemistry, focused upon the explanation of macroscopic physical and chemical properties in light of the basic components and structure of molecules, the defining activity of physical chemistry according to prominent physical chemists such as Hans Landolt and Arnold Eucken.[50] Neither topic remains a fundamental aspect of physical chemistry today in large part because disciplinary developments in the 1930s belied chemists' expectations. Colloid chemistry became a specialty with myriad potential applications but little claim to general significance, while physics subsumed almost entirely the investigation of atomic structure. This was hardly the outcome Haber and Freundlich envisaged, and the apparent diversity of the investigations they promoted at the Institute during the Weimar era reflected neither confusion nor pure expediency but rather their conviction that research into a broad range of phenomena, some marginal to mainstream chemistry at the time, could reveal general principles of demonstrable import to all branches of chemistry, pure and applied.[51]

Colloid chemistry, the unfailing mainstay of the Freundlich department, traces its roots back to the discovery in the 1860s by the Scottish chemist Thomas Graham that certain aqueous solutions passed through a semi-permeable membrane only with difficulty, if at all. These solutions he termed "colloids," to distinguish them from "crystalloids" which passed with ease through such a membrane. Graham presented his research on colloids as a fascinating aspect of the specialized field of solution chemistry. A more modern understanding of colloids, extending the definition of colloids to any substance in which one chemical compound is microscopically distributed through another, regardless of the phases of either substance, and focusing on the role of surface forces in colloid behavior, only took shape some four decades later. It was largely the result of a campaign, spearheaded by Wolfgang Ostwald, son of the famous founder of physical chemistry, to establish colloid research as an independent and fundamental chemical discipline. Colloid chemistry was primarily an experimental endeavor, and in addition to redefining colloids with respect to both their material properties and their scientific significance, the younger Ostwald and his allies developed and refined a host of instruments and techniques to advance the new discipline, many of which, such as the ultramicroscope and electrophoresis, remain familiar to chemists today.[52]

Herbert Freundlich belonged to the vanguard of the campaign for colloid chemistry. He completed his doctorate under Wilhelm Ostwald in Leipzig just a year

49 Haber to Harnack, 9 June 1923, MGPA Abt. Va, Rep. 5, Nr. 1916.
50 Landolt, *Antrittsrede*. Eucken, *Grundriss*.
51 Haber, *Zeitlalter*.
52 Cf. Ede, *Rise*.

before Wolfgang Ostwald did the same. Two years later, Freundlich began publishing in the flagship journal of the field, *Kolloid Zeitschrift*. The journal was edited by the younger Ostwald for over three decades and would host the overwhelming majority of publications by Freundlich and his collaborators during that period. Like many colloid chemists, Freundlich also emphasized the importance of colloids to biology, and he supported research in this vein in his department of the Institute. However, even when researching the properties of biological compounds, Freundlich framed his experiments as investigations into the general principles of colloid behavior, fully in keeping with the tenor of the Ostwald campaign.[53]

Freundlich first made a name for himself in capillary and adsorption chemistry, investigating the thermodynamics of liquid-solid and gas-solid interfaces and the differences between various adsorption phenomena, in modern terms the distinction between chemisorption and physisorption. Following the broader definition of colloids proffered by Ostwald, Freundlich saw this research as bearing directly upon colloid chemistry in that it aimed at clarifying the general principles of surface interactions. In his research on adsorption, Freundlich relied heavily upon the earlier work of Josiah Willard Gibbs to develop a quantitative account of the phenomena early researchers had only described qualitatively. Perhaps the most enduring legacies of this research are the Freundlich isotherm, relating gas adsorption to pressure at a given temperature, and Freundlich's textbook on capillary chemistry, which went through four editions and remained a standard work in the field for many years.[54] However, of greater immediate import for Freundlich's arrival at the Institute was his choice of experimental systems. Freundlich studied the adsorption of non-electrolytes and weak electrolytes on activated charcoal, which became a key component in German gas mask filters – hence his invitation to leave behind the Technical College at Braunschweig and join Haber's Institute in 1916.

Immediately after the war, Freundlich had two primary assistants, Alexander Nathansohn and Hans Kautsky. Nathansohn was a rarity, an independent chemist who developed practical industrial applications. He worked with Freundlich on "wet metallurgy," i.e. the application of knowledge concerning metal solutions and colloids to refining procedures, and developed a patented method for separating lead from tin in mixed ores with high sulfur content.[55] However, Nathansohn also encouraged, if not sparked, Freundlich's interest in the interaction of colloids with light. In addition to an article with Nathansohn on photochemical reactions in colloids, Freundlich published an article on the electrocapillarity of colored solutions in collaboration with Marie Wreschner.[56] Freundlich and Wreschner included industrially significant dyestuffs in their research, but they maintained a focus on the general phenomena of electrocapillarity, using a technique Haber developed in collaboration with Klemensiewicz while at Karlsruhe to measure the potential

---

53 On biography, cf. Donnan, *Freundlich*; Reitstötter, *Freundlich*.
54 Freundlich, *Chemistry*.
55 Nathansohn, *Rohstoffe*.
56 Freundlich, Nathansohn, *Lichtempfindlichkeit*.

Fig. 2.17. *Herbert Freundlich (1880–1941), 1931.*

of glass electrodes then relating this potential to the migration of dyestuff ions in solution.[57]

The articles with Nathansohn and Wreschner were just the beginnings of research on electrocapillarity and the optical properties of colloids in the Freundlich group. Freundlich himself went on to attempt a general thermodynamic account of electrocapillarity, similar to his earlier work on adsorption. Further exploration of the optical properties of colloids devolved to Hans Kautsky, who was soon joined by Hans Zocher. Kautsky had begun work under Freundlich during the war, before completing his doctorate.[58] In the years immediately after the war, in addition to finishing his degree, Kautsky helped develop techniques for measuring the photosensitivity of colloid solutions, focusing on inorganic solutions. Biochemistry, however, would become the best-acknowledged beneficiary of Kautsky's research. An extension of the experimental techniques Kautsky began developing at the Institute led, after Kautsky's departure for Heidelberg, to the first recognition of the Hirsch-Kautsky effect, the characteristic quenching of chlorophyll fluorescence. Hans Zocher's research, on the other hand, focused on optical and magnetic anisotropy in colloid systems, including the streaming or flow birefringence identified by Georg Quincke at the turn of the century. Zocher's recognition of the relation between the asymmetry of colloid particles, the anisotropy of their structure when stressed and their birefringence is often cited as one of the earliest steps toward the development of liquid crystal technologies.[59] As with Kautsky and luminescence, Zocher published his most widely cited works in this field after his

---

57 Freundlich, Wreschner, *Elektrokapillarkurve*.
58 Cf. Jaenicke, *Kautsky*.
59 Cf. Demus, *Zocher*.

departure from Haber's institute, in Zocher's case while a professor at the Technical University in Prague during the 1930s. However, both lines of research clearly originated at the Institute and relied, in their formative stages, upon the work of Institute colleagues. Kautsky drew, in particular, upon the work of Haber and Zisch on luminescence,[60] while Zocher had immediate access to X-ray crystallographic studies of colloid structure, as well as a rudimentary theory of the structure of particulate colloids by Eugene Wigner and Andor Szegvari.[61]

Upon his arrival in the summer of 1921, Georg Ettisch added a new facet to research in the Freundlich department. Ettisch, who remained at the Institute until forced to leave in 1933, embodied Freundlich's belief in the biological significance of colloid chemistry. Whereas Freundlich restricted himself to using biologically significant compounds in studies of general colloid phenomena, such as adsorption or coagulation, Ettisch endeavored to relate these general phenomena to specific biological functions.[62] This was a widespread field of research during the 1920s, but its pursuit often led to tensions between chemists and medical researchers, as chemists frequently showed little respect for clinical research protocols and posited simple mechanisms to explain complex biological phenomena on the basis of exclusively laboratory research. Ettisch, however, remained relatively conservative in his speculations on biological function and took seriously the knowledge of medical researchers, completing his own habilitation in the medical faculty of the Berlin University in 1929 and becoming a welcome expert on the physical chemistry of colloids amongst medical researchers. Nevertheless, even after Ettisch took charge of his own working group, the Institute did not rank among the key centers for biochemical colloid research. Ettisch and his collaborators published careful studies of coagulation and the colloid behavior of blood serum and similar substances, but their research was neither pivotal to the realization that proteins and other substances vital to the structure and sustenance of life were in fact macro-molecules, rather than colloids of small particles, nor did it contribute to the refinement of methods for analyzing biological substances, such as electrophoresis and ultracentrifugation, that developed in the context of colloid chemistry but survived the transition to macro-molecular models.[63]

Instead, what colloid research at the Institute became widely known for, in addition to photochemistry and capillary chemistry, was the study of thixotropy, the reversible conversion of a semi-rigid gel to a fluid sol through shaking, stirring or similar prolonged exposure to shearing forces. Before his arrival at the Institute, Freundlich had studied the coagulation of hydrophobic sols upon the addition of electrolytes, searching both for ways to quantify the phenomenon and to explain it on the basis of known intermolecular forces. In the first years after the war, Freundlich returned to the study of coagulation only in passing, but he took a renewed interest in the topic after two junior researchers in his department, Emma

60 Haber, Zisch, *Anregung*.
61 Szegevari, Wigner, *Stäbchensolen*.
62 Cf. Rürup, Schüring, *Schicksale*, p. 187–188.
63 About this development cf. e.g. Deichmann, *Molecular*.

Fig. 2.18.   *Herbert Freundlich's Colloid Chemistry Department, end of the 1920s.*

Schalek and Andor Szegvari, observed reversible sol-gel transitions in iron oxide colloids.[64] Freundlich coined the term thixotropy to describe the phenomenon and made the study of reversible transition phenomena a mainstay of his research for the remainder of his career. Initially, the iron oxide observations inspired several short-lived investigations that met with varying degrees of success, including a study of the structures of iron and aluminum hydroxides, both ingredients in thixotropic colloids, carried out by Johann Böhm, after 1925 the husband of Emma Schalek.[65] Freundlich, however, soon embarked on a more systematic study of coagulation times and possible mechanisms for thixotropy. In 1928, Karl Söllner joined Freundlich in this research, and together they extended the study of thixotropy from transitions induced by mechanical stress to those induced by ultrasound.[66] This research did not lead Freundlich and Söllner to fundamental new insights into cohesive forces, but it did lay the groundwork for our present understanding of a phenomenon vital to numerous industrial products including solder pastes and certain adhesives.

Members of the Freundlich group pursued each of these topics both individually and through short- and long-term collaborations. Partnerships spanning several years, like that of Freundlich and Söllner or Kautsky and Zocher, were

---

64 Schalek, Szegvari, *Eisenoxydgallerten.*
65 Böhm, *Aluminiumhydroxide.*
66 Freundlich, *Thixotropie.* Freundlich, Rogowski, Söllner, *Utraschallswellen.*

interspersed with one-time collaborations that sometimes included researchers from other departments or even other research institutes in Berlin, such as the Szegvari and Wigner study of the electrical behavior of particulate colloids. This suggests permeable boundaries between departments, as well as open circulation of research results within each department, and marked freedom for individual researchers to pursue at least short-term investigations as they saw fit. It is a pattern that one also finds within the Physical Chemistry Department under Haber. Although, in the case of the Physical Chemistry Department overlaps with physics research at the Institute were even more pronounced.

Physical and Theoretical Chemistry

Haber spent some portion of the limited time he personally felt able to dedicate to pure research in the first years after the war developing new models of the structure of solids. The best known result of these efforts is the Haber-Born cycle, a thermodynamic analysis of the formation of ionic crystals into component steps corresponding to energies (ionization energy, electron affinity, etc.) whose sum gives the total energy of formation of the crystal; it was frequently used to calculate lattice energies, the one step in the cycle that cannot generally be measured directly. The Haber-Born cooperation resulted unexpectedly from the frequent trips Max Born made to Berlin to visit James Franck. Born was initially wary of Haber, as Born was opposed to chemical warfare, but Haber managed to win his confidence and arrange a brief collaboration with long-lasting results.[67] The Born-Haber research had clear antecedents in the work of Born and Alfred Landé on crystal lattice energies, but Haber too made roughly contemporaneous attempts to calculate macroscopic crystal properties on the basis of atomic-scale models, albeit with a focus upon metal structure and less lasting success.[68]

Haber's definitive contribution to the research direction of the Institute in the early 1920s, however, would be an article with Walter Zisch on the emission of light during combustion reactions. Haber and Zisch studied the spectra emitted by ordinary flames, as well as those produced during reactions of alkali and halogen gases, in particular sodium and chlorine. They observed that these reactions emitted light that could not be ascribed to the heat of the reaction and that could be seen even when the reaction mixture was not hot enough to glow visibly according to the laws of blackbody radiation. To control the reaction temperature Haber and Zisch allowed the gases to react only at very low pressures, creating "highly-dilute flames." They posited that the process that produced this anomalous light, and which they took to be representative of chemiluminescence in general, was the inverse of a photochemical reaction. That is, they argued that light was emitted

---

67 Einstein, Born, *Briefwechsel*, p. 40. Born, *Life*, p. 261 f.
68 Sauer, *Superconductivity*, p. 193–195.

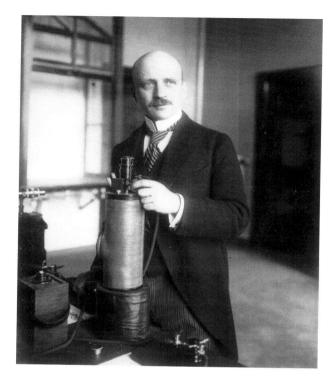

Fig. 2.19. *Fritz Haber in the laboratory, 1922.*

after the electrons of one or more of the reactants or products were excited by the reaction and then returned to their ground states, emitting light through a process involving fluorescence. The phenomenon was markedly more complex than simple excited fluorescence or classical photochemical reactions because the nuclei and the electrons could interact in ways not fully understood at the time, and the emitted light could originate from an electronic transition in any one of the reactants, intermediates or products present in the reaction vessel. Hence, Haber and Zisch argued, one could not expect a one-to-one correspondence between the number of molecules of product formed and the number of light quanta emitted, as one might expect from a straightforward inversion of the photochemical decomposition mechanism, but one could very likely detect unstable intermediates and gain insights into reaction mechanisms by studying chemiluminescent spectra. Both the complexity of the reaction mechanisms suggested by the Haber and Zisch article and the possible insights one might gain into reaction mechanisms by careful spectral observations became launching points for new lines of research at the Institute, including the research of Hans Kautsky on chemiluminescence in colloids and the experimental work of the Polanyi group on reaction kinetics in the gas phase, as well as Haber's own return to combustion research after 1926.

Concurrent with his first chemiluminescence research, Haber also advised dissertations by Fritz Schmid and Hans Lehrecke that laid the foundations for the gold from seawater project.[69] Both doctoral candidates wrote on new methods for determining the gold content of highly-dilute solutions, with Schmid focusing specifically on seawater. Haber's exact role in these researches is somewhat unclear, as is the division of labor between Haber and Johannes Jaenicke, who became head of Department in 1922. Haber was clearly on board during the ocean voyages of the *Hansa* and the *Württemberg* in 1923 and later wrote an article detailing the results of the project.[70] But the key insights regarding new methods of filtration and cupellation that informed experimental work during the ocean voyages appeared first in doctoral dissertations, as did the realization by Isaak Rabinowitsch that ostensibly large variations in gold content between seawater samples were actually the result of gold "pollution" and could be eliminated through more careful handling and storage procedures. Haber, however, showed a clear practical understanding of Rabinowitsch's insights in a series of articles from 1926 debunking claims by German and Japanese chemists that they had transformed mercury into gold.[71]

Several independent strands of research emerged within Haber's "physical chemistry" department while Haber was underway with Department M. One of the most prestigious began with the arrival of Karl-Friedrich Bonhoeffer at the Institute in 1923 and his subsequent investigations of the chemistry of activated, i.e. atomic, hydrogen. Given that atomic hydrogen was the only substance for which physicists felt they might have an acceptable quantum theoretical model at the time, it was a topic with manifest, if nebulous, potential for bridging physics and chemistry. From the outset, Bonhoeffer took advantage of the expertise in low-pressure gas chemistry available at the Institute. As he branched out to research on simple hydrogen containing compounds, he also benefited from its spectroscopic facilities and from the assistance of Ladislaus Farkas, with whom he established a connection between the diffuse bands in the electronic spectra of ammonia and predissocciation and then interpreted the bands' widths in terms of the energy-time uncertainty relation.[72] Bonhoeffer's research took a distinct turn, however, after the arrival of a new collaborator, Paul Harteck, in 1928. Harteck had habilitated under Max Bodenstein in Berlin and then spent two years working as an assistant to Arnold Eucken in Breslau, at the time a center for low-temperature specific heat experiments.

Bonhoeffer and Harteck set out together to confirm the existence of two recently-posited, distinct forms of molecular hydrogen: ortho-hydrogen, with nuclear spins oriented parallel to one another, and para-hydrogen, with mutually opposing nuclear spins. Ever since Arnold Eucken first measured the specific heat of hydrogen gas at low temperatures in 1912, its anomalous temperature dependence had posed a challenge for quantum theories of specific heat. In 1927, working in close correspondence, Werner Heisenberg and Friedrich Hund, each

69 Lehrecke, *Lösungen*; Schmid, *Gold*.
70 Haber, *Gold*
71 Haber, Jaenicke, Matthias, *Darstellung*.
72 Bonhoeffer, Farkas, *Predissociation*.

Fig. 2.20. *The Institute's hydrogen team, circa 1930. Left to right: Ladislaus Farkas, Paul Harteck, Adalbert Farkas, Karl Friedrich Bonhoeffer.*

independently published articles in which they suggested that hydrogen existed in distinct ortho- and para-forms. Furthermore, they argued that these two forms should exist in a ratio of 3 to 1 at high temperatures, but, as Hund pointed out, should contribute differently to the specific heat of the gas. Later that year, David Dennison combined these insights into a theory that fully accounted for the observed specific heat of hydrogen, and included the premises that at low temperatures the para- rather than the ortho-form would be favored but the transition between the two states would be slow.[73] It was this transition that would allow Bonhoeffer and Harteck to test the new theory, but it was a challenging task, Harteck's Breslau training in low-temperature methods notwithstanding. In a letter from 28 October 1928, Bonhoeffer wrote:

> we have set our minds upon an experiment that should show that ordinary hydrogen ... is a mixture, as the theorists believe ... but it isn't working at present, and I have lost half my hair to the futile drudgery[74]

---

73 Cf. Gearhard, *Hydrogen.*
74 PHP, 1 : 1.

practice rather than scientific method. Polanyi argued against the notion that science was a dispassionate pursuit of impersonal facts, whose procedures could be made fully explicit, but strove to do so without undermining the general validity of scientific knowledge. The cornerstones of his work in these fields were the concepts he presented in his books *Personal Knowledge* (1958) and *The Tacit Dimension* (1966). Both "personal knowledge" and "tacit knowledge," like Thomas Kuhn's concept of a "paradigm," have since found audiences well beyond science studies. Both concepts also shared roots in deeply held views on economics and science policy that Polanyi had developed through years of discussions with other scientists and economists, including his older brother Karl, a noted socialist political economist, and through multiple personal encounters with political unrest and with Communism, as well as through his own work as a scientist. In this last respect Polanyi was distinct from many of his colleagues in philosophy of science, few of whom had pursued distinguished research careers in the natural sciences.

When Polanyi left Dahlem for the University of Manchester in 1933, he was at the height of his career in chemistry, but by the end of the 1930s his focus was clearly shifting toward social topics. Economics was his first non-chemical interest. He produced a short film in 1938 extolling the virtues of a Keynesian market economy, partly inspired by a trip to the USSR in 1935, during which he was appalled by the failures of the centrally planned economy. The film was a limited success, but he continued to develop and refine his ideas on economics, and to date Polanyi remains a widely-recognized figure amongst neo-Keynesian economists. In the Soviet Union, he also encountered the Lamarckian, state-supported doctrine of Trofim Lysenko, which rejected the modern concepts of Mendelian genetics, and witnessed the persecution of the doctrine's critics. This, as well as his opposition to Soviet central planning, helped motivate Polanyi's departure from science to the philosophy of science and science policy. As historian Mary Jo Nye has pointed out, the ideal organization of science that Polanyi contrasted with the shortcomings of central planning closely resembles his own descriptions of his experiences as a researcher at Haber's Institute.

In reaction to Polanyi's new research interests, the University of Manchester created a personal chair for him in "Social Studies" in 1948. In 1951 the University of Chicago offered him a position in social philosophy, which he accepted, but which he was unable to assume because the U.S. government would not grant him a visa, ironically because they were suspicious of his contact with communists. In 1959, he became a Senior Research Fellow at Oxford, but held the position for only two years before reaching mandatory retirement age. After retirement, he traveled widely and gave guest lectures at a number of prestigious universities.

To date there are three scholarly societies dedicated to studying the life and work of Michael Polanyi: the Michael Polanyi Liberal Philosophical Association in Budapest, which publishes the journal *Polanyiana*; the Polanyi Society in the United States, which publishes *Tradition and Discovery*; and the Society for Post-Critical and Personalist Studies in Britain, which publishes *Appraisal*. He and his wife Magda had two sons. The elder, George Polanyi, became a successful economist. The younger, John Charles Polanyi, shared the 1986 Nobel Prize in Chemistry for his "contributions concerning the dynamics of chemical elementary processes."

Fig. 2.22. *Henry Eyring (1901–1981), circa 1942.*

studying simple reaction rates through chemiluminescence.[81] On the theoretical side, Polanyi collaborated with Eugene Wigner, who completed a dissertation at the Technical University in Charlottenburg under Polanyi's supervision in 1925 on the "Formation and Decay of Molecules, Statistical Mechanics, and Reaction Velocities."[82] Wigner first joined Polanyi while the latter was still at the Institute for Fiber Chemistry, where Wigner also worked with Hermann Mark. After the completion of his degree and a brief return to Hungary, Wigner then rejoined the Fiber Institute as an assistant to Karl Weissenberg, pursuing research on the use of symmetry groups in crystal structure analysis, but he continued to aid Polanyi in kinetics research as well. The Polanyi group was also aided in their theory endeavors by Fritz London,[83] whose many contributions to physics included developing one of the first quantum mechanical accounts of the covalent chemical bond, in collaboration with Walter Heitler. These two men, together with Hans Beutler, brought to the reaction kinetics research a facility with quantum mechanics, especially its more advanced mathematical methods, that Polanyi lacked.

In his 1920 article on reaction kinetics, Polanyi noted that existing kinetic theories could not be quite correct, as the ratio of forward to backward reaction rates failed to yield the equilibrium constants obtained on the basis of thermodynamics. In 1925, in a rejoinder to a paper in which Max Born and James Franck argued that it would be nearly impossible for a collision of molecules to incite chemical reactions, Polanyi and Wigner managed to resolve the discrepancy between forward and reverse reaction rates for the case of two-body capture and its reverse,

81 Beutler, Polanyi, *Reaktionsleuchten.*
82 On biography, cf. Mehra, *Wigner.*
83 For biography, cf. Gavroglu, *London.*

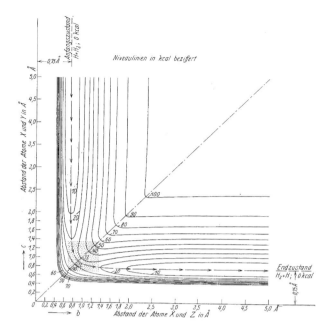

Fig. 2.23. *Potential energy surface of the H + H₂ ⇔ H₂ + H reaction for a collinear collision geometry as reported by Henry Eyring and Michael Polanyi in 1931.[86]*

one-body decay.[84] This laid the groundwork for the later Breit-Wigner formula (1936), which describes the kinetics of both molecular and nuclear near-resonant collisions. The Polanyi-Wigner article was followed by a series of articles from Hans Beutler examining in detail the quantum mechanics of the collision of gas particles and the resulting excitation of electrons, composed in part in collaboration with Polanyi and in part with Eugene Rabinowitsch. Starting from a general study of atomic collisions by Hartmut Kallmann and Fritz London, Beutler and his colleagues treated in detail the kinds of collisions they thought most likely to contribute to chemical reactions and chemiluminescence.[85] But the most enduring theoretical achievement of the Polanyi group would depend pivotally upon the help of one of the Institutes many visiting foreign scholars.

In 1929, Henry Eyring, arrived at the Institute with a National Research Council Fellowship to pursue research with Polanyi. Eyring was a graduate of the University of California, Berkeley, but had spent the preceding two years doing research and teaching at the University of Wisconsin. Eyring initially aided Polanyi with highly-dilute flame experiments, but Polanyi soon invited Eyring to assist him in applying recent work by London on the dynamics of chemical reactions, in particular the quantum mechanics of the making and breaking of chemical bonds,

---

84 Polanyi, Wigner, *Molekülen*.
85 Beutler, Rabinowitsch, *Drehimpuls*; Beutler, Rabinowitsch, *Energieanreicherung*.
86 Ibid.

to the problem of chemical activation energy. In their landmark 1931 article on the subject, Polanyi and Eyring relied upon the hydrogen research of Bonhoeffer, Harteck and the Farkas brothers. In particular, they employed as a model system for their more detailed calculations the simplest chemical exchange reaction, $H + H_2 \Leftrightarrow H_2 + H$, which Adalbert Farkas had posited as the mechanism for the interconversion of ortho- and para-hydrogen. Together Polanyi and Eyring established a visual metaphor for understanding the process of making and breaking of chemical bonds which, for thermal and hyperthermal reactions, persists until today:

> the chemical initial and final state are two minima of energy which are separated by a chain of energy mountains. [...] Among all possible paths [across the mountains], the reaction path is the one which leads over the lowest pass, whose energy elevation determines the activation energy of the reaction.[87]

To understand the progress of a reaction one needs only to imagine a ball representing the configuration of the nuclei of the constituent atoms rolling along the potential energy surface containing this "chain of mountains," following a path determined by the disposal of energy of the reaction, comprised of translational, vibrational and rotational components. Implicit in this model from the outset was a separation of the nuclear and electronic motions now referred to as the Born-Oppenheimer approximation. Initially, Polanyi and Eyring also relied upon calculations by Fritz London to define the energy surfaces appropriate to the reactions they were studying, but they soon began using spectroscopic results to refine their estimates of electronic energies, creating an innovative "semi-empirical" method.[88]

Pelzer and Wigner, both members of the Polanyi group, then combined these semi-empirical potential energy surfaces with considerations from statistical mechanics into an analysis of reaction rates that would form the starting point for "transition state" (Polanyi) or "activated complex" (Eyring) theory. However, neither Polanyi nor Eyring published his first article on transition state theory until after they had both departed Haber's Institute, Polanyi for Manchester and Eyring for Berkley then Princeton.

The reaction mechanism studies of Tokyo-University-trained physical chemist Juro Horiuti present a similar case. In the spring of 1933, Horiuti joined the Polanyi group from Arnold Eucken's laboratory, where he had been doing research on Raman spectra. When he arrived at the Institute he began research with Polanyi on heavy water. Their collaboration then turned to studies of hydrogen exchange reactions and eventually resulted in the first descriptions of the Horiuti-Polanyi mechanism. This research marked the beginning of Horiuti's life-long interest in catalysis and electrochemistry, and to date the Horiuti-Polanyi mechanism remains a preferred model for the hydrogenation of hydrocarbons at solid surfaces. But the culmination of this research came only after Horiuti followed Polanyi to

---

87 Polanyi, Eyring, *Gasreaktionen*, p. 280.
88 Nye, *Tools*.

Fig. 2.24. *Polanyi's Physical Chemistry Department, 1933.*

Manchester in August of 1933.[89] As with the discovery of the Hirsch-Kautsky effect and Zocher's work on the optical properties of liquid crystals, the experiences of Eyring and Horiuti provide excellent examples of the opportunity the Institute provided some researchers to launch lines of research that would continue to shape their careers well after they left Dahlem.

Spectroscopy and Quantum Physics

Reaction mechanism research also marks the third instance, along with chemiluminscence and combustion research, in which spectroscopy was indispensable to a line of chemical research at the Institute. Spectroscopy and the related phenomenon of dispersion were primarily the purview of physicists during the first decades of the 20th century and were central to the development of quantum theory and quantum mechanics.[90] However, on the basis of examples such as Max Bodenstein's groundbreaking research into photochemistry and chain reactions, many chemists, particularly physical chemists, also believed spectroscopy could offer insights into key chemical questions. In Copenhagen, the physical chemist

---

89 Cf. Hirota, *Horiuti*; Horiuti, *Early Days*.
90 Cf. Friedrich, Hoffmann, *Quantum Physics*.

Fig. 2.25. *James Franck (1882–1964), circa 1925.*

Niels Bjerrum established one of the most famous and innovative laboratories for the study of band spectra, and in France and the United States, leading physical chemists Jean Perrin and Gilbert N. Lewis generalized from existing photochemical studies to the ill-fated "radiation hypothesis." Its central premise, that all chemical reactions depend upon the absorption of light of specific frequencies, encouraged chemists to attend more carefully to the optical phenomena accompanying chemical reactions. Haber neither took up spectroscopic research himself, nor became an enthusiast of the Perrin-Lewis radiation hypothesis,[91] but spectroscopy formed the backbone of the physics research undertaken at his institute during the Weimar era and embodied his standing conviction that chemists could benefit from attending more closely to quantum theory. Not coincidentally, this also led to research being done at the Institute that contributed directly to scientists understanding of and confidence in quantum theory.

Under the direction of James Franck, in the years immediately following the war, Gustav Hertz, Erich Einsporn, Walter Grotrian and Paul Knipping concentrated primarily on the careful measurement of absorption spectra and ionization energies and the correlation of these measurements with the Bohr-Sommerfeld model of the atom.[92] It was an extension of a line of research Franck and Hertz had begun while at the Berlin University before the war. There they devised the highly-admired "Franck-Hertz experiment," which demonstrated that electron collisions with mercury vapor atoms were elastic only up to a certain threshold energy, and that beyond this threshold inelastic collisions led to ionization and electronic excitation of the atoms.[93] The specific ionization and excitation energies they

---

91   Daniels, *Radiation.*
92   Cf. Lemmerich, *Sturm.*
93   Franck, Hertz, *Zusammenstöße.*

observed corresponded with predictions based on Niels Bohr's quantum model of the atom, providing it strong experimental support. Though performed at an ostensibly chemical institute, their post-war efforts were similarly central to quantum theory, as their careful measurements of spectra and ionization energies "enabled the confirmation of Bohr's theory to a high degree of precision."[94] Their results also formed part of the basis for several later investigations at the Institute, including, as already discussed, the pivotal 1922 article by Haber and Zisch, as well as the work of Hans Beutler and others on the quantum mechanics of atomic collisions. In this respect, the focus of the Franck group on the spectra of mercury vapor was particularly important, as their exemplary results encouraged later researchers at the Institute to choose mercury vapor as a model system. In collaboration with Fritz Reiche, the "house theorist" at the Institute, Franck also published an article on helium and para-helium describing cases in which excited electrons were unable to return directly to the ground states, an early example of so-called "forbidden transitions."[95]

In the interregnum between James Franck and Rudolf Ladenburg, Paul Knipping published retrospective articles on the discovery and practice of X-ray diffraction and descriptions of a new apparatus for ionization measurements.[96] More importantly, the X-ray apparatus and spectroscopic equipment at the Institute offered members of the Physics Department the opportunity to branch out into research on the Compton effect, the shift in the wavelength of X-rays caused by inelastic scattering from an electron. The discovery of the Compton effect in 1923 caused quite a stir in the physics community, as it provided strong support for the particulate nature of X-rays and, by extension, of light. However, Compton's results proved somewhat difficult to replicate. Noted Harvard X-ray physicist William Duane tried and failed, but Hartmut Kallmann of the Physics Department in collaboration with Hermann Mark, an expert on X-ray analysis at the Fiber Institute, were able to reproduce the phenomenon and to make careful measurements of the relationship between the scattering angle and the shift in wavelength, which they published in a 1925 issue of *Die Naturwissenschaften*.[97] Kallmann also performed a more rigorous physical analysis of the ongoing research at the Institute into the excitation of gas spectra through chemical reactions, e.g. Haber and Zisch, for which he found an audience in *Zeitschrift für Physik*. Still, there is nothing to indicate that these were aspects of a department-wide program of research, and Kallmann, though nominally attached to the Physics Department, pursued his research interests essentially independently even before he became head of his own working group in 1928.[98]

A bit further afield from the Physics Department proper, but still in keeping with the interest in spectroscopy and atomic physics at the Institute, were the enduring

---

94 Franck, Einsporn, *Quecksilberdampfes*.
95 Franck, Reiche, *Helium*.
96 Knipping, *Zehn Jahre*. Knipping, *Registrierapparat*.
97 Kallmann, Mark, *Comptoneffekte*.
98 Cf. Wolff, *Kallmann*.

Fig. 2.26.   *Eugene Wigner (1902–1995) with Werner Heisenberg (left), 1928.*

contributions of Eugen Wigner to quantum theory. In 1926–1927, while still dividing his research efforts between the Herzog and Haber Institutes, Wigner became the first scientist to employ group-theoretical considerations in the interpretation of the selection rules of atomic spectroscopy. He accomplished this by analyzing the transformation properties of energy eigenstates of a system with respect to operations that leave the system physically unchanged, e.g. spatial rotations, mirror inversions, exchange of identical electrons. Wigner had developed his skills with group theory and symmetry transformations while working with Weissenberg on crystallography, a field in which these mathematical tools had been commonplace ever since Evgraph Fedorov and Arthur Schoenflies characterized, at the end of the 19th century, the 230 "space groups" describing all possible crystal symmetries. More recently, space groups had taken on an even greater significance in crystallography thanks to publications by Ralph Wyckoff (1922), and by Kathleen Yardsley (later Dame Lonsdale) and William Astbury (1924) that systematically linked each of the space groups to specific "characteristic absences" in X-ray diffraction patterns, making them indispensable tools for X-ray structure analyses. Symmetry groups, however, had not yet made similar inroads into other branches of physics, and many physicists were initially hostile to their importation into quantum theory, even referring to them as the *Gruppenpest*, the group plague.[99]

---

99 Cf. Chayut, *Periphery*; Borrelli, *Selection*.

Nevertheless, the encounter Wigner arranged between group theory and the old-quantum-theoretical notion of selection rules had a profound and long-lasting impact on quantum theory.[100] The interpretation of spectroscopic evidence in terms of stationary states and selection rules had been an important conceptual model for spectroscopists working with the old quantum theory. The connection between selection rules and group theory endowed quantum mechanics with a new type of symmetry argument, in which selection rules, rather than conservation laws, were regarded as the observable signature of an underlying physical symmetry. Interpreting experimental data in terms of selection rules, therefore, led to a redefinition of the traditional conserved quantities, notably angular momentum.[101] In fact, it was Wigner who, in a short paper published in 1927, drew attention to the new, quantum form of conservation laws, articulating what is today referred to as the quantum version of Noether's theorem. Wigner noted that in quantum mechanics one was only allowed to ask about the probability distribution of the values of physical quantities and concluded:

> It is therefore necessary to formulate also the laws of conservation in this sense. They will then have the form, for example: The probability that the energy will have the value E does not change with time.[102]

When asked in the early 1930s by Max von Laue what group theoretical result derived so far was the most important, Wigner replied: the explanation of the Laporte rule (the concept of parity) and the quantum theory of vector addition (angular momentum). Partly in recognition of the power of these new theoretical tools, Wigner would receive the 1963 Physics Noble prize "for his contributions to the theory of the atomic nucleus and the elementary particles, particularly through the discovery and application of fundamental symmetry principles."

While Wigner was taking the first steps to integrate group theory and quantum theory, a new line of experimental research was developing within the Physics Department, as Rudolf Ladenburg, in collaboration with Hans Kopfermann and later Agathe Carst, undertook a series of experiments intended to test the new quantum theory of dispersion. Dispersion played a central role in the development of quantum theory in general, and in the formulation of the matrix mechanics by Werner Heisenberg in particular,[103] and during his time in Breslau, Ladenburg had made important contributions to the transformation of classical dispersion theory into its quantum counterpart. In Breslau, Ladenburg was assisted in this research by his friend and colleague Fritz Reiche, who had been previously affiliated with Haber's Institute. In Dahlem, the task fell primarily to Hans Kopfermann, who arrived at the Institute in 1926, immediately after completing his habilitation in Göttingen under James Franck.[104] Ladenburg and Kopfermann (and later Carst)

---

100 Borrelli, Friedrich, *Wigner*.
101 Borrelli, *Selection Rules*.
102 Wigner, *Erhaltungssätze*, p. 381.
103 Jansen, Duncan, *Umdeutung*.
104 Schlüpmann, *Kopfermann*; Lieb, *Kopfermann*.

Fig. 2.27. *Rudolf Ladenburg (1882–1952), circa 1930.*

compared the predictions of the latest versions of the quantum theory with novel experiments on dispersion in excited gases, and as Haber reported to a meeting of the Prussian Academy in June of 1926:

> Using the method of interference bands, anomalous dispersion was confirmed, and in some cases measured, in several lines of the He, Ne, Hg and H [spectra], when the gases were excited by a continuous current. On the basis of the quantum theoretical dispersion formula of Ladenburg and Kramers and the F-summation rule of Reiche-Thomas these measurements were used to determine the probabilities of various quantum transitions, as well as the number of atoms in the excited states and their dependence upon current and the temperature and pressure of the gas.[105]

The continuation of this line of research also led to a series of articles published in *Zeitschrift für Physik* between 1928 and 1930, in which they presented the first evidence of "negative dispersion," what physicists now call stimulated emission.[106] According to the quantum dispersion theory, as formulated independently by Ralph Kronig (1926) and Hendrik Kramers (1927), one could create a sample material that, when illuminated by light of the appropriate frequency actually emitted more light of that frequency than it absorbed. This is now recognized as the crucial phenomenon behind the operation of lasers, and some historians of science have even argued that with just a bit more luck Ladenburg and Kopfermann might have observed the first laser pulse.[107] After Günther Wolfsohn took over as Ladenburg's assistant in 1930, Kopfermann turned his attention to the hyperfine structure of atomic spectra. His investigations of the spectra of different isotopes during the

105 Haber, *Dispersion*.
106 Ladenburg, Kopfermann, *Negative Dispersion*.
107 Brown, Pike, *Optics*.

Fig. 2.28. *Hans Kopfermann (1895–1963), circa 1928.*

following year contributed to the discovery of the "isotopic shift," the effects of the nucleus on the energy of the surrounding electrons. Exploring the properties of the nucleus through its interactions with electron orbitals would develop into Kopfermann's research specialty when he later moved to professorships at Kiel and then at Heidelberg, and he wrote one of the standard early works on the topic, *Kernmomente*.[108] As with Paul Harteck, this clear move toward nuclear studies laid the groundwork for his later participation in uranium research under the National Socialist regime.

Of course, not all of the research undertaken at the Institute during its first "golden era" found such a grand fate. There were myriad short-lived offshoots of the researches discussed here and numerous collaborators whose research has not been covered in detail. Visiting Japanese scholars supported by the new stipends established in the 1920s helped Freundlich with several projects in colloid chemistry. Karl Weissenberg, inventor of the eponymous X-ray goniometer, contributed X-ray analyses of carbon containing crystals that helped convince organic chemists of the value of the new technique. Michael Polanyi and Erika Cremer even extended kinetics and reaction mechanism research at the Institute to the study of reactions at gas-solid interfaces, broaching topics to which scientists at the Institute would return independently in the 1980s. But these offshoots were no more likely than the researches previously discussed to fall outside the central research foci of the Institute, meaning not just "colloid chemistry" or "atomic structure" in general, but the more specific topics within these fields around which publications from the Institute clustered, such as surface energy, coagulation, photochemistry, reaction mechanisms and combustion reactions.

---

108 Kopfermann, *Kernmomente*.

The interconnectedness of the researches at the Institute notwithstanding, at times their topics of investigation may appear to have led members of the Institute beyond the bounds of physical chemistry, at least as we now see them. But disciplinary boundaries, however clear they may appear in pedagogy and funding practices, are often difficult to discern in ongoing research.[109] As discussed earlier, spectroscopy was, in very particular respects, of clear interest to chemists in the 1920s; it remains so today. But judging in precisely which respects spectroscopy is "chemistry" and in which respects "physics" requires a thorough understanding of the technique, if it is possible at all. Haber, like many of his contemporaries in physical chemistry, did not feel it necessary to wait for and obey such clear disciplinary demarcations when choosing topics of research, and in the context of early 20th-century physical chemistry this strategy often led to highly-regarded results, as illustrated in the case of the Haber Institute by the prestigious universities offering faculty positions to its long-standing members, including Breslau (Reiche), Hamburg (Harteck), Harvard and Frankfurt (Bonhoeffer), Manchester (Polanyi), Minneapolis (Freundlich) and Princeton (Ladenburg).

---

109 Cf. Barkan, *Nernst*.

Fig. 3.2. *Max Planck and Minister of the Interior Wilhelm Frick, 1933.*

of Jews – some useful to mankind and others worthless... one must observe the distinctions."[7]

Fritz Haber, however, was not quite so ready simply to obey the expectations, or rather orders, of the General Administration and the Ministry. He wanted to remain in command for the time being because, as he expressed to Schmidt-Ott on 21 April 1933,

> The structure of this institute is the most important of my personal responsibilities as its head. Should current circumstances render this structure unsustainable, apparently because it has become disadvantageous to the Kaiser Wilhelm Society and the institute, which I have led since its foundation, I consider it my duty as director to see through the required reorganization myself because I know best those aspects [of its structure] that are important for science and for the personnel and am best placed to make arrangements with the General Administration.[8]

Both department heads, Herbert Freundlich and Michael Polanyi, requested retirement on the same day Haber wrote this letter to Schmidt-Ott. Soon Haber too had decided, after much difficult deliberation, that he would resign from his post as institute director once "his duties were complete;" with the inevitable consequence, according to Haber, "that the many individual questions related to the

---

7 Planck, *Besuch* 1947, p. 143.
8 MPGA Abt. I, 1a, Nr.541/3, Bl. 2.

required reorganization would resolve themselves." On April 30, Haber asked the ministry that he be allowed to retire on 30 September 1933 – the date by which the Civil Service Law had to be implemented. In his request, which would later become famous, he clearly stated that he had always sought colleagues for his institute according to qualification and character, a stance from which he was both unable and unwilling to retreat, despite the new regulations.

The leading figures of the KWG, specifically President Max Planck and General Director Friedrich Glum, continued to try and convince Haber to change his mind. They not only regretted losing a highly-valued and internationally-renowned colleague but also saw Haber's demonstrative act as damaging to the KWG. They also feared that, through Haber's retirement, their influence on a central institute within the Society would be severely compromised. But Haber was not inclined to change his mind, and the decision became irreversible after Rust delivered two speeches in which he polemicized Haber's request for release, making Haber a *persona non grata* for the NS leadership.

The resignation of Haber and his department heads was regarded as a kind of signal within the top ranks of the National Socialists and beyond. It was seen as a protest against the arbitrary racist directives and against the National Socialist state more broadly. It was all the more striking because all three men, as veterans of the First World War and long-standing civil servants, were exempt from the Civil Service Law and legally could have remained in their positions. Concerns for the future of his institute and its staff occupied Haber's remaining weeks in office. His last official acts were efforts to limit as much as possible the impact of the Civil Service Law and to exploit the remaining legal framework so that he could provide the best possible provisions for staff members who had lost their positions. Having been instructed by the General Administration on 9 July 1933 to dismiss seven colleagues before the summer break, he tried to win exemptions for some of them on grounds of hardship, as well as attempting to arrange new positions for them. Rita Cracauer, the "soul of the Institute," who had been Haber's secretary for many years; Hartmut Kallman, Haber's "right hand," and Irene Sackur, daughter of Otto Sackur, were particularly difficult cases for Haber. Cracauer had made a career of her work at the Institute and was otherwise without means; moreover, her brother had been killed in the First World War. Haber felt a special obligation to Irene Sackur, who had only joined the institute in 1931, because of the fatal injury her father had suffered while working at the Institute early in the First World War. In the end, Haber's petitions made little headway in the Ministry, and 10,000 RM were taken from the Haber Fund to help the dismissed colleagues and somewhat alleviate their financial distress. Rita Cracauer, following Haber's emigration, looked after his remaining property in Berlin. She subsequently emigrated via Great Britain to Palestine. Irene Sackur was able to stay on at the Institute for a short time but was soon after denounced and had to leave in the fall of 1933. She also emigrated to Palestine in the mid-1930s. It was easier for the world-renowned scientists. Freundlich first emigrated to London and worked at University College beginning in 1934. He then went to Minneapolis in the United States, where

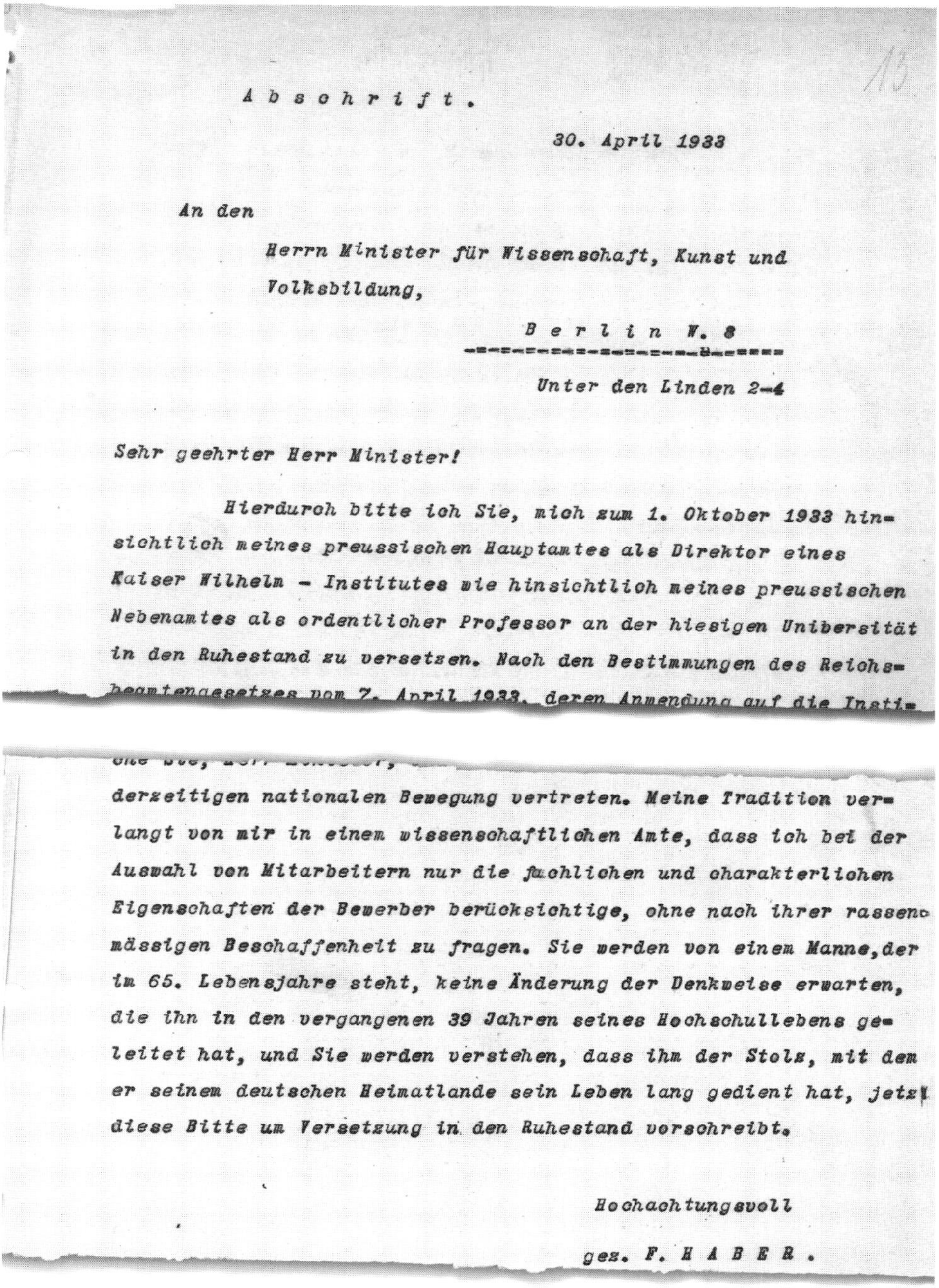

A b s c h r i f t .

13

30. April 1933

An den

Herrn Minister für Wissenschaft, Kunst und
Volksbildung,

B e r l i n W. 8

Unter den Linden 2–4

Sehr geehrter Herr Minister!

Hierdurch bitte ich Sie, mich zum 1. Oktober 1933 hinsichtlich meines preussischen Hauptamtes als Direktor eines
Kaiser Wilhelm – Institutes wie hinsichtlich meines preussischen
Nebenamtes als ordentlicher Professor an der hiesigen Universität
in den Ruhestand zu versetzen. Nach den Bestimmungen des Reichs-
beamtengesetzes von 7. April 1933, deren Anwendung auf die Insti-

derzeitigen nationalen Bewegung vertreten. Meine Tradition ver-
langt von mir in einem wissenschaftlichen Amte, dass ich bei der
Auswahl von Mitarbeitern nur die fachlichen und charakterlichen
Eigenschaften der Bewerber berücksichtige, ohne nach ihrer rassen-
mässigen Beschaffenheit zu fragen. Sie werden von einem Manne, der
im 65. Lebensjahre steht, keine Änderung der Denkweise erwarten,
die ihn in den vergangenen 39 Jahren seines Hochschullebens ge-
leitet hat, und Sie werden verstehen, dass ihm der Stolz, mit dem
er seinem deutschen Heimatlande sein Leben lang gedient hat, jetzt
diese Bitte um Versetzung in den Ruhestand vorschreibt.

Hochachtungsvoll

gez. F. H A B E R .

Fig. 3.3.   *Resignation letter from Haber to Berhard Rust, 30 April 1933.*

Fig. 3.4.   *Farewell gathering in the Institute garden, July 1933. First row standing on the right: Friedrich Epstein; seated from the right: Hartmut Kallmann, Michael Polanyi, Fritz Haber; seated in front of Haber: Rita Cracauer; two chairs left of Haber: Herbert Freundlich; far left, seated on the ground: Karl Klein, glassblower.*

he worked as Research Professor of Colloid Chemistry until his untimely death in 1941. He was accompanied in his travels and aided in his research by his colleague Karl Söllner, who remained in the United States after Freundlich died.[9] Polanyi became Professor of Physical Chemistry in Manchester, a position he had been offered in 1932 but turned down in mid-January 1933. In Manchester, he initially continued his research into the transition state of chemical reactions, in particular the catalytic conversion of hydrogen, but he later turned his attention to philosophy and the social sciences.[10] Ladislaus Farkas emigrated to Palestine in 1934, where he established the Department of Physical Chemistry at the Hebrew University of Jerusalem with his brother Adalbert.[11] Despite serious efforts, Hartmut Kallman did not succeed in emigrating. As a result of his "privileged mixed marriage," he became an attendant at I.G. Farbenindustrie and AEG, and survived the Holocaust in Germany.[12]

When Haber first left Germany, in September 1933, he went to Cambridge, where he was given the opportunity to work in the laboratories of William Pope; however, this hospitality had limited consequences. Although he did manage to perform some experiments on catalytic decomposition of hydrogen peroxide with the help of his Dahlem assistant, Joseph Weiss, his exile was more than anything

---

9   Reitstötter, *Freundlich*.
10 Nye, *Polanyi*.
11 Pallo, *Farkas*.
12 Wolff, *Kallmann*. As for Kallmann's "privileged mixed marriage" status, cf. his own account in a letter to Federal President T. Heuss of 4(?) January 1954, MPGA, PA Kallmann.

Fig. 3.9. *Peter Adolf Thiessen surrounded by colleagues at the Bunsen Congress in Düsseldorf, 1938. Left to right: Erich Hückel, Peter Debye, Peter Adolf Thiessen, Klaus Clusius and Hans-Heinrich Frank.*

Another strong point of his research was conductometric analysis and its practical applications; though, his publications in this field appeared primarily in handbooks rather than journals.[25] This reflected well Jander's position in the scientific community; he did not present fundamentally new results but was widely recognized as an expert in his field. Jander may also have continued to pursue military research that he had begun in secret while still in Göttingen. In any case, when Thiessen took over in 1935, he stated explicitly in the plans he provided the Board of Directors that the Institute "would, for the moment, be dedicated to those tasks that Ministry of Defense indicated to him were urgent." This allowed Thiessen to continue the research carried out under Jander since the beginning of 1934.

On taking over the institute, Thiessen invested most of his efforts into establishing a modern, technical infrastructure, and in 1936, he initiated a comprehensive reorganization of the Institute, which took place against the backdrop of far-reaching changes in the organization of research into chemical warfare by the Army Ordnance Office.[26] As a result of the Third Reich's public rearmament policy, the Ordnance Office was making plans for its own centralized

25 E.g. Jander, *Maßanalyse*. Jander, Pfundt, *Leitfähigkeitsreaktionen*.
26 Cf. Schmaltz, *Kampfstoff-Forschung*, p. 100–124.

Peter Adolf Thiessen (1899–1990)

Peter Adolf Thiessen was born into the family of a landowner in Schweidnitz, Silesia. After his high-school (Gymnasium) graduation he volunteered for service in WWI. He studied Chemistry in Breslau, Freiburg, Greifswald and Göttingen, where he graduated in 1923 with a thesis on colloidal gold under Richard von Zsigmondy, for whom he subsequently became an assistant. After his habilitation in 1926 he stayed on as a privatdozent and after 1932 as Extraordinarius Professor for Physical Chemistry; nevertheless, he failed to succeed his mentor as Ordinarius Professor for Inorganic Chemistry. Thiessen's engrossment in the National Socialist movement dates back to his Göttingen time – already in 1922 he was a member of the NSDAP and the SA, whose local structures he helped to shape. However, in order not to imperil his university career, he took a break from his party affiliation later in the 1920s, then re-activated it instantly once the Nazis ascended to power. In 1933 he moved to Berlin, where he took the post of department head at and, in 1935, Director of the KWI for Physical Chemistry and Electrochemistry. Parallel to his scientific activities he contributed to the Nazi transformation of the German university-education system as a rapporteur in the Ministry of Education. He remained actively involved in Nazi science policy even after he withdrew from this post in the mid-1930s – after 1937 in the capacity of Chemistry Division leader of the newly created Research Council (Reichsforschungsrat). His involvement in research policy, the prestigious directorship of the KWI, membership in the Berlin Academy (since 1939), chairmanship of the Bunsen Society (1942–45) as well as other positions made Thiessen into one of the most influential and powerful scientists and science managers in the Third Reich.

In 1945, Thiessen accepted an offer to work for the Soviets, so that in the following years, together with other German "specialists" at a secret research institute in the Caucasus, he contributed to the Soviet nuclear bomb project. In 1956, he arrived in his new homeland, the GDR, where he was able to start a new career at the Academy of Sciences (which had expelled him in 1945, because of his Nazi past) as Director of the Central Institute for Physical Chemistry and professor at Humboldt University in Berlin. In the GDR as well Thiessen brushed shoulders with political power: from 1957 to 1965 he was the chairman of the Research Council and from 1960 to 1963 member of the State Council (Government) of the GDR. Highly decorated, Thiessen died at age 90 in (East) Berlin.

Fig. 3.12. *Framing of the new X-ray building, 1938.*

testament to Gert Molière's close ties to the Institute, his younger brother Kurt completed a dissertation at the Institute in 1939 on the influence of absorption on the diffraction of electron beams. His approach resembled that of his older brother in that he sought to formulate specific laws of electron diffraction based on quantum mechanics and building on works by Laue and by Hans Bethe. To provide an experimental comparison, the younger Molière applied the theory to zinc blende, whose crystal structure had been determined over a decade earlier. By April of 1940 the Institute had gained another expert in applied mathematics, Bernard Baule. Baule had been a student of Hilbert in Göttingen and had also been active in the Catholic student associations in Graz.[38] He was apparently released from "protective custody" thanks to Thiessen,[39] and while at the Institute, he performed thermodynamic calculations and helped to analyze X-ray diffraction patterns.

Beginning in 1934, August Winkel directed an independent Department for Colloid Chemistry, through which he furthered the research on aerosols, smokes and fogs that he had previously pursued under Jander. Winkel emphasized the relevance of research on these colloidal systems to meteorology and to occupational health, e.g. protection from inhaled particulates through smoke and dust filters. He also noted, markedly more reticently, the possible military applications

---

38 Vortrag Kurt Überreiter, Berlin, 2. Juli 1981, MPGA Abt. VII/2 Tonträger, Überreiter T 135 1/2. Weigand, *Die Technische Hochschule Graz.*

39 Aktennotiz Walther Forstmann an Ernst Telschow, 8 May 1940, MPGA Abt. I, 1a, Nr. 1175.

Fig. 3.13. *August Winkel using the ultramicroscope, circa 1938.*

of such research, which, in addition to smoke screens, included distribution of poison gases, which were generally aerosols of toxic liquids. It is very likely that the Winkel department also carried out researches on filters to guard against chemical weapons and on filter-breaking compounds. Filtration research focused mainly on adsorption filters and porous materials, but also extended to industrial electro-filtering, which was important in the recovery of scarce raw and manufacturing materials. As part of this research, Hans Witzmann sought to establish a systematic basis for characterizing various filters. He established elementary laws of filtration by conducting experiments on model substances and introduced the K_z value as a characteristic measure of filter effectiveness.[40] Since aerosols, smokes and fogs are difficult to produce and last only a short time, establishing such standards required extensive experimentation. Modern analytical methods were used to study particles as small as 0.1 µm; amongst the main methods were light absorption spectroscopy, X-ray crystallography and, most important of all, ultramicroscopy. Later studies used electron diffraction and electron microscopy. Conductivity measurements also ranked among the key analytical tools in Winkel's department, as they had in his teacher Jander's. In this respect the polarography techniques developed by Jaroslav Heyrovský in Prague were particularly important. These techniques were also used in analyses of the structures of organic molecules, in which differences in the reduction potential of individual functional

---

40 Witzmann, *Elementarvorgänge*.

groups, e.g. keto, carbonyl or carboxyl, were related to the chemical structures surrounding the group.[41]

Dietrich Beischer, who began his scientific career in Winkel's department, developed pioneering electron microscopy techniques. At the close of the 1930s, he began to focus on the preparation of samples for electron microscopy, work he pursued first in collaboration with Friedrich Krause,[42] a colleague of Ruska's at the High Voltage Institute in Neubabelsberg, and later in collaboration with Manfred von Ardenne. In 1938, Beischer gained access to the first, prototype raster electron microscope, housed in Ardenne's laboratory in nearby Lichterfelde. This groundbreaking device allowed certain structural details of catalysts, plastics, carbon blacks, metal oxide smokes and rubbers, including buna synthetic rubber, to be observed for the first time.[43] Building on the optical microscopy results of Hermann Staudinger and previous X-ray analyses of fibers, Beischer made visible the thread-like molecular bundles of high polymers and placed them under mechanical stress in order to better understand how changes in their microscopic structure affected their macroscopic properties. In 1941, Beischer was appointed to the University of Strasbourg, where he pursued further investigations using electron microscopy. Electron microscopy research also continued at the KWI for Physical Chemistry and Electrochemistry, making use of the electromagnetic "*Über*-Microscope" developed by Siemens based on a design by Ruska beginning in 1939 and an electrostatic microscope designed by Hans Mahl at AEG beginning in 1940.

Ernst Jenckel, who completed his doctorate under Tammann in Göttingen in 1932 and received his habilitation under Schenck in Münster, arrived at the Institute in 1935. His research focused on analyses of the structures of glasses, polymers and alloys. With respect to metal alloys, he systematically investigated changes induced in their mechanical properties by changes in the structures of their solid solutions; his research on glasses similarly sought to relate microscopic structure to physical properties and also included investigations of glass-like synthetics. Jenckel developed a new line of research around synthetics, borrowing concepts and categories from the study of glasses and molten materials. The research carried out under Jenckel was also of pronounced strategic importance, as attested to by the fact that he was assigned the task in 1938 of developing his own Four-Year-Plan Institute. This new institute was housed at the KWI for Physical Chemistry and Electrochemistry before being moved to the TH Aachen in 1941.

Kurt Überreiter joined the Institute in 1937 as a doctoral student under Jenckel. He grew to be an important member of the Institute staff and carried out essential research into the structure and rigidity of plastics. Überreiter developed two fundamental concepts in early plastics research. First he characterized the glass-like

---

41 Proske, Winkel, *Über die elektrolytische Reduktion.*
42 Beischer, Krause, *Elektronenmikroskop.*
43 E.g. Ardenne, Beischer, *Katalysatoren.*

Fig. 3.14.   *Ernst Jenkel, left, with a colleague in front of an air liquefier, circa 1938.*

state of rubber and synthetic resins that arises when the cooling process occurs quicker than the relaxation time of the polymer melt as a "liquid of fixed structure." He identified the transition to this state via a kink in the volume versus temperature curve, and in the case of rubber, discovered that the transition occurred at the unexpectedly low temperature of around −65 °C.[44] He attributed the elasticity of rubber to this low transition temperature. In general, he explained his observations concerning this state through the limited mobility of the individual segments of chains and nets of molecules. The second fundamental concept Überreiter made clear was the distinction between internal and external plasticization of polymers.[45] Internal plasticizers become part of the polymer chain or net, preventing rigid crystallization; whereas, external plasticizers are not chemically bound to the polymer and function somewhat analogously to solvents. During the war, Überreiter began research with a more applied bent, including analyses of the quality and effectiveness of fillers, such as zinc oxide or carbon black, which is particularly important for rubber production. His findings indicated that surface structure of particles was of considerable significance, which lent even greater importance to electron microscopy and strengthened the tie to the work of Schoon and Beischer.

44 Überreiter, *Kautschuk und Kunstharze*.
45 Überreiter, *Weichmachung*.

In 1936, a department for research in organic chemistry was created under the leadership of Arthur Lüttringhaus, a student of Windaus who specialized in the synthesis and analysis of elongated cyclic molecules such as cyclic ethers and thioethers. Through experiments with ring-closing reactions, he became the first to determine specific bond angles between carbon, oxygen and sulfur atoms in these compounds using a classic chemical approach.[46] At the Institute, Lüttringhaus had the advantage of readily-available, high-quality diffraction analyses to which he could compare his results, thus combining crystallographic and synthetic methods of structure analysis. After Lüttringhaus accepted the offer of an associate professorship for organic chemistry at the University of Greifswald in 1940, Alfred Pongratz took over the department. Pongratz had previously pursued his research at the University of Graz and was particularly interested in gas-phase, catalytic oxidation.

In 1940, a department for special colloid chemistry research was established under the direction of Otto Kratky,[47] who would remain at the Institute only three years before accepting a position at the Technical University in Prague. Kratky was an expert in X-ray crystallography; while working under Mark in Vienna, he had developed small-angle scattering into an effective method for determining the structure of very large aggregates of similar molecules. He focused on the structural analysis of macromolecular materials, both natural and synthetic, and worked extensively with cellulose obtained from viscose, which was important to the industrial production of artificial silk, with cellulose film and with spun rayon. Researchers in his department were able to discern the attachment of the thread-like molecules of cellulose to structures that were in part crystalline and flake-like and in part amorphous and tangled. This complemented the previously discussed results from the Thiessen group concerning micelles of long-chain molecules. Kratky also investigated protein structures. In 1942, he and his colleague Aurelie Sekora were the first to confirm the spherical structure of chymotrypsin.[48] Franz Seelich headed another, biology-oriented, working group. Seelich had worked previously at the Pasteur Institute in Paris. In 1927 he became an assistant professor in Kiel. At the Institute after 1941, he investigated the effect of anesthetics on cells and tissues.

Germany's troubled international relations notwithstanding, the Institute hardly operated in isolation. In addition to its military contacts, it had ties to industrial research institutes, which often provided samples for structural analysis, as well as to other scientific research facilities. In the early 1940s, for example, Rudolf Kohlhaas moved from Dahlem to Leuna, one of I.G. Farben's most modern research, development and production sites. Around 1941, Georg Richard Otto Schultze, a specialist in mineral oil technologies and hydrogenation with professional experience in the U.S., who served as an assistant at the University of Berlin

---

46 E.g. Hauschild, Lüttringhaus, *Valenzwinkelstudien*.
47 Ausschnitt Deutsche Allg. Zeitung, 12 September 1940, MPGA 1. Abt., 1a, Nr.1175.
48 Kratky, Sekora, *Röntgenstrahlen*.

and who later became a full professor at Technical University Braunschweig, conducted chemical processing research at the Institute as a guest scientist. There was also wide-ranging cooperation with the other Dahlem institutes. For example, Georg Graue, a former student of Hahn at the neighboring KWI for Chemistry introduced a radioactive emission technique developed there to the Thiessen Institute, and in 1938, Otto Hahn left the Institute a hardness testing machine based on Caldwell's design. Kratky cooperated with members of the KWI for Biochemistry, Hans Hermann and Hans Friedrich-Freksa, on the analysis of protein structures,[49] and Theodor Schoon produced electron microscope images of iron oxides in Dahlem that were simultaneously being studied using X-ray analysis at the Inorganic Chemical Institute at TU Stuttgart.[50]

During the course of the war, the number of projects of immediate military relevance increased. In 1938, at the request of Mentzel and the SS, Thiessen set up a small, predominantly-secret military department for Eugen Weber. The department occupied a single laboratory at the Institute and probably investigated microfilming techniques and counterfeit money and documents.[51] Winkel and Witzmann, both closely linked to Jander, were also members of the SS since 1931 and 1932 respectively and rose through its ranks.[52] The research group led by Anton Bartels, whom Thiessen had brought to the Institute from Leuna in 1941, carried out another project with clear military relevance. It focused on friction across bearings and lubrication, and it set up its instruments for measuring bearing wear in the main building of the KWI for Chemistry. In November 1943, at Thiessen's invitation, an external branch of the Army Ordnance Office headed by Horst Böhme, whose offices had been destroyed by bombing, moved into the KWI for Physical Chemistry and Electrochemistry. Shortly before the end of the war, Böhme and his office relocated to Hesse. Böhme went on to become Professor for Pharmaceutical and Organic Chemistry in Marburg in 1946.[53] Robert Haul, who had joined Winkel's staff in 1937 and completed his doctorate at the TH Berlin in 1938, was another direct contributor to the war effort. Beginning in 1944, he headed an external department of the KWI for Physical Chemistry and Electrochemistry based at an explosives institute of the Weapons Office in Prague. There he appears to have followed the "Dahlem style" in employing the latest in analytical equipment and techniques. Haul was clearly in contact with Jaroslav Heyrovský and had access to his polarographic equipment, which registered its output on an oscilloscope,[54] and which Haul used to determine the reduction potentials of fuels and explosives.[55]

---

49 Kratky, Sekora, Weber, *Kleinwinkelinterferenzen.* Friedrich-Freksa, Kratky, Sekora, *Röntgeninterferenzen.*

50 Fricke, Schoon, Schröder, *Umwandlungsreihe.*

51 Interview Klaus Thiessen.

52 On Winkel, cf. Deichmann, *Flüchten,* p. 232–233 and p. 545 and Schmaltz, *Kampfstoff-Forschung,* p. 33, p. 78, p. 106–118 and p. 127–134. On Witzmann, cf. Deichmann, *Flüchten,* p. 545 and Schmaltz, *Kampfstoff-Forschung,* p. 108–109, p. 133 and p. 137.

53 Schmaltz, *Kampstoff-Forschung,* p. 118–123.

54 Podaný, *Heyrovský,* p. 547.

55 Haul, Scholz, *Grenzflächen-Reaktionen,* p. 232–234.

Table 3.2. *Publications of the technical (workshop) staff of the KWI for Physical Chemistry and Electrochemistry during the Third Reich:*

Wilhelm Ulfert	Zerspanung des Stahles mit 18 % Chrom und 8 % Nickel mit Werkzeugen aus Silberstahl, Stahl und Eisen 55 (1935).
Erich Franke	Eine vielseitig verwendbare Vakuumkammer für Röntgenfeinstrukturaufnahmen, Zeitschrift für physikalische Chemie 31 (1936).
Walter Spatz	Verbesserung der Mikrobürette, Chemische Fabrik 9 (1936).
Kurt Hauschild	Fraktionierte Vakuumdestillation fester Substanzen, Chemische Fabrik 10 (1937).
Karl Klein	Verbesserte Quecksilberreinigung, Chemische Fabrik 10 (1937).
Karl Klein	Über einen neuartigen Thermoregler, Zeitschrift für Instrumentenkunde 59 (1939).
Wilhelm Ulfert	Ein Präzisions-Schlagzahn, Feinmechanik und Präzision 48 (1940).
Karl Klein	Feinfraktionierkolonne ganz aus Glas unter Verwendung von Mehrkammerrohren, Zeitschrift für physikalische Chemie A 189 (1941).
Wolfgang Srocke	Ein Winkel-Krauskopf, ein verstellbarer Drehstahl-Halter, Feinmechanik und Präzision 50 (1942).
Wilhelm Ulfert	Ankörn- und Zentriergerät, Feinmechanik und Präzision 50 (1942).
Wilhelm Ulfert	Verstellbare Bohrvorrichtung, Feinmechanik und Präzision 50 (1942).

technical staff were not only thanked in articles, they were even credited as co-authors. Moreover, they started publishing short articles under their own names in technical journals and magazines explaining improvements they had made to instruments, apparatus or tools. Again, this had no precedent at the Institute. Wilhelm Ulfert, who had been employed as an apprentice mechanic under Haber and had become the head of the NSDAP organizational cell in 1933,[63] took full advantage of these new opportunities. But Kurt Hauschild, who, unlike Ulfert, was not a National Socialist and who continued working at the Institute until the 1970s, also benefited greatly from the new publishing arrangements.

Although the technical members of staff did not receive pay raises or managerial positions, the importance of their work was at least more directly acknowledged than in the past. This was not just a reflection of their significance to the Institute, but was also in line with the pseudo-egalitarian ideology of the NS regime. Apparently, Thiessen himself was a paradigm of this approach. He was widely considered to be decent and even-handed, and he protected individual employees who had come into conflict with other parts of the NS system. For example, thanks to their active participation in the National Socialist power structure, he and his leading men were able to have a number of employees exempted from military service.

---

63 Aktenvermerk Ernst Telschow 29 August 1933, MPGA I. Abt. 1a, Nr. 1168.

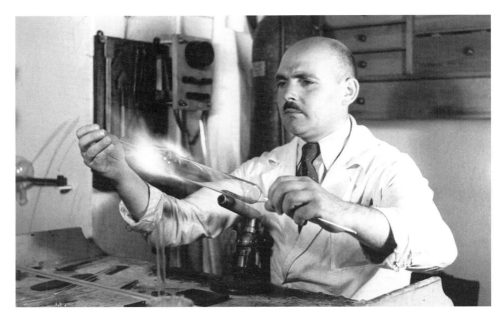

Fig. 3.17. *Glassblower Karl Klein, circa 1938.*

In 1937, Thiessen took on principal responsibility for chemistry and organic materials at the Reich Research Council (*Reichsforschungsrat*).[64] Later physical chemistry was added to his purview. The Research Council was responsible for funding decisions within the German Scientific Research Association (DFG), which replaced the Emergency Asssociation of German Science (NG) in 1935 and was headed by Mentzel from 1936 onwards. In his work for the Research Council, Thiessen not only contributed significantly to realizing the broad directives provided by National Socialist science and research policies, in particular to promoting research in the field of chemistry oriented toward the Third Reich's self-sufficiency policy, but also advocated close links between theory and practice in chemistry. Very much in the spirit of Wilhelm Ostwald, he repeatedly described physical chemistry as "general chemistry," in that it provides the foundations for all chemical specialties. A more distinguishing characteristic of Thiessen's science policy activities and of his guidance of the Institute, particularly in the context of the NS regime, was his support for a certain degree of creative freedom, which he even defended against anti-intellectualist attacks.[65] However, he argued that the *raison d'être* of basic research was its relevance to practical applications, and in this respect his views anticipated the linear model of the relationship between science and technology that was later prominently advocated by Vannevar Bush in the U.S.

---

64 Schmaltz, *Kampfstoff-Forschung*, p. 125 ff.
65 Thiessen, *Physikalische Chemie*.

of the German higher education system and research community. The German Research University was officially founded in May of 1948 but would be dissolved just five years later.

As one of its first official acts, the Board of Trustees of the DFH named Karl Friedrich Bonhoeffer Director of the Kaiser Wilhelm Institute for Physical Chemistry and Electrochemistry. Bonhoeffer, who began his scientific career at the Institute under Haber, had been appointed *ordinarius* Professor for Physical Chemistry at Berlin University in 1946 and, since then, had reestablished close contacts with the Institute in Dahlem. At the beginning of 1947, he received an informal inquiry from Göttingen asking whether he would be willing to take over the Institute in Berlin, and although the Central Administration for Public Education approved him for the post in April, Bonhoeffer was not offered the appointment until a year later. The cause of this delay was the inchoate state of the Kaiser Wilhelm/Max Planck Society, which first officially formed in Göttingen in February of 1948 within the British and American Bizone. The Society's central task was the preservation or resurrection of the institutes and research stations of the earlier Kaiser Wilhelm Society, and they sent feelers out to Berlin in an attempt to include in their plans the remaining institutes of the KWG located there, in particular the Institute for Physical Chemistry and Electrochemistry. However, there were also considerable efforts being made in Göttingen at the time to found a new institute for physical chemistry that might initially make use of those parts of the Dahlem institute that had ended up in the West during the last months of the war, such as Joachim Stauff's group and the institute's library. The plan succeeded, and as a result, on 1 February 1949 Bonhoeffer took charge of the MPI for Physical Chemistry in Göttingen. This meant he had to commute between posts in Berlin and Göttingen. He increasingly viewed his visits to Berlin as an onerous duty since the focus of his scientific activities had rapidly shifted toward the building up of the new institute in Göttingen, and besides, as he confided to his friend Paul Harteck, he enjoyed "tremendously the comforts of the West."

In Bonhoeffer, the Institute had once again found a director who conformed to the standards of the Max Planck Society and, moreover, enjoyed the trust of the Göttingen central MPG administration. Bonhoeffer also served as an emissary to the Berlin municipal authorities in negotiations concerning the integration of the Dahlem institutes into the MPG. These discussions occurred as the German Research University struggled for its existence. It had given rise to no noteworthy activities and, early in the 1950s, began descending into an ever deeper crisis. Part of the reason for the crisis was that the Freie Universität Berlin was established in Dahlem early in 1948, as a counterbalance to Humboldt University (formerly *Friedrich-Wilhelms Universität* or Berlin University) in (East) Berlin, and began attracting progressively more of the limited American interest and support. Moreover, with the founding of the Federal Republic of Germany, interest of the supporting German Lands (states of the German Federal Republic) in financing such a grand research institution, located in remote (West) Berlin and operating in parallel to the MPG, waned. In contrast, after the Königsteiner agreement of

Fig. 4.4.   *Robert Havemann at a protest in West-Berlin against the nuclear arms race, 18 July 1950.*

1949, the financing of the MPG, and hence its ongoing existence, was fundamentally secure, leading to stronger consideration being given the reintegration of the Berlin institutes, especially the KWI for Physical Chemistry and Electrochemistry.

The dismantling of the DFH was accompanied by ever increasing restrictions on Havemann's sphere of influence. When the Berlin municipal authorities backed out of financing specialized research groups at the Research University early in 1949, financing for Havemann's department, which was wholly funded by the municipality, also came up for negotiation. In the end, Havemann remaining at the Institute was assured through the efforts of Bonhoeffer, who not only drew attention to Havemann's former service, but also stated that he expected "from [Havemann's] scientific work an increase in the renown of the Institute."[9] In the preceding years, Havemann had resumed his pre-war research activities and occupied himself with a variety of problems in colloid chemistry, especially those relating to the physical chemistry and biochemistry of proteins. Nevertheless Havemann continued to see himself as a public citizen. He made no secret of his communist convictions and campaigned for political organizations and initiatives of the German Democratic Republic (GDR), which had been established in the fall of 1949. He served as a delegate to the People's Parliament (*Volkskammer*), the representative assembly of the GDR, as well as remaining involved in ongoing daily political developments. In

---

9 Bonhoeffer to Verwaltung des KWI, 25 March 1949. ARHG.

the heated political atmosphere of the border town of (West) Berlin this gave rise to suspicions and numerous accusations from the political authorities. This led, in July of 1949, to his removal from his post as administrative head of the Berlin Kaiser Wilhelm Institutes through the actions of the American authorities; he had supposedly "enabled illegal scientific research, permitted and encouraged it," in violation of ACC Law 25. Then, early in February of 1950, Havemann voiced his opposition to the American atomic weapons policy in the (East) Berlin newspaper *Neues Deutschland*, the central outlet of the state communist party, in an article that was as much a policy critique as a polemic and that included an explanation, in layman's terms, of the principles behind the operation of the hydrogen bomb; it attracted extensive and sensational commentary from the media in the West. Some even voiced the suspicion that Havemann might be an "atomic spy," and could have been in contact with Klaus Fuchs, the German-born Soviet spy who had been unmasked just a few weeks earlier. The West Berlin authorities and the American occupying forces backing them used this affair as their opportunity to remove Havemann without notice from his position as department head at the Institute. Bonhoeffer genuinely tried to negotiate a compromise between the two parties, but both the Senate's decision and Havemann's convictions were unwavering. From then on Havemann lived and worked in the GDR. For some time he was Professor at Humboldt University, as well as a sincere partisan of the prevailing East-German Communist Party (*Sozialistische Einheitspartei Deutschlands*, SED), but in the 1960s, he turned into one of the most prominent dissidents in the GDR.

Bonhoeffer finally took his leave from the Dahlem institute in 1951, and on April 1, Max von Laue took over the Institute, the incorporation of which into the MPG appeared to be only a question of time. Laue was the 1914 physics Nobel laureate and had studied at Berlin University, completing his doctorate in 1903 under Max Planck; he continued to work at the University, side by side with his revered mentor, from 1920 to 1944. Hence, Laue felt a special bond with Berlin and with the Kaiser Wilhelm and Max Planck Societies.[10] Furthermore, in spite of his 71 years, he proved to be a dynamic science manager. He deployed his considerable international renown to the benefit of the Institute and he worked intently toward the incorporation of the Berlin Kaiser Wilhelm Institutes into the Göttingen-based Max Planck Society. Initially, hard feelings and contested financial arrangements in Dahlem hindered his pursuit of this goal. More specifically, the MPG felt itself to be the direct legal successor of the KWG and entitled to compensatory damages from the Freie Universität for the use of former KWG properties. In addition, the Berlin scientific administration did not want to lose its influence over the Dahlem institutes and was worried by the possibility of an exodus of qualified scientists from Berlin; Bonhoeffer's move from Berlin to Göttingen had hardly allayed these fears.

On the other hand, the Berlin Blockade and the aggressive Soviet stance toward West Berlin and Germany had made it clear that West Berlin would only be able to survive if the city sector maintained the closest possible ties to the Federal

---

10 Cf. Hoffmann, *Laue*.

Fig. 4.5. *Karl Friedrich Bonhoeffer at the unveiling of the Haber commemorative plaque on 9 December 1952.*

Republic. The nurturing of cultural and scientific relations, in part through the founding of appropriate institutions, played a key role in this strategy, alongside the more obvious adoption of the economic, financial and legal systems of the Federal Republic. The governing Mayor of West Berlin, Ernst Reuter, was one of the most avid advocates of this policy, which would remain a mainstay of West Berlin politics until German reunification. The integration of the Dahlem Kaiser Wilhelm Institutes into the Max Planck Society was, therefore, significant not only to the institutes themselves but also to the city of Berlin, for which it had political and symbolic value as a step toward the West. Hence, the Berlin Senate was prepared to meet the MPG more than halfway, in particular with respect to questions of endowment and concessions in the property dispute with the Freie Universität. The MPG would henceforth be entitled categorically to the property rights of the KWG in Dahlem, which rapidly brought to a close negotiations with the Freie Universität over damage provisions for the buildings it occupied. At the same time, the MPG entered into negotiations with the German Research University and the Berlin Senate and showed its willingness to reciprocate with the establishment of an administrative center in Berlin. The original demand from Reuter that the MPG move its headquarters back to Berlin was untenable in the eyes of the MPG and also appeared unrealistic from a political standpoint, given the vulnerable island position of Berlin in East Germany.

Still the name of the Dahlem institute, i.e. "Institute for Physical Chemistry and Electrochemistry," presented something of a quandary since there was now an MPI for Physical Chemistry in Göttingen.[11] Following a proposal from Max von Laue, the name was changed to "Fritz Haber Institute of the Max Planck Society in the framework of the German Research University," thus avoiding any potential ambiguity; as of 1953 this became simply Fritz Haber Institute of the Max Planck Society (*Fritz-Haber-Institut der Max-Planck-Gesellschaft*). Laue further highlighted the legacy of Haber at the Institute by organizing a grand celebration for Haber's birthday, during the course of which a bronze commemorative plaque was unveiled in the stairwell of the Institute's main building. Bonhoeffer was the keynote speaker at the event, and he strove to make good for every word that he had been forbidden to utter at the 1935 commemoration by Rust's ministerial ban.

On 29 January 1952, the MPG and the DFH concluded a cooperation agreement, but possible terms for the reintegration of the Berlin institutes into the MPG were still unclear. Intensive negotiations followed between the General Administration of the MPG, the DFH, the Berlin Senate and the "Berlin commission" of the MPG, which had been established for this purpose and placed under the direction of Bonhoeffer. On 8 January 1953, the Berlin commission suggested the admission to the MPG of sundry institutes of the DFH, most of which were either previous KWG institutes or departments of former institutes. On Febuary 4, in the buildings of Warburg's institute, the Board of Trustees of the DFH, members of the MPG General Administration and representatives of the city of Berlin concluded a corresponding contract, which was then ratified by the senate of the MPG at a general assembly in Harnack House on 20 May 1953.[12] The Berlin commission had fulfilled its duties and was dissolved with the approval of Mayor Ernst Reuter. In his address to the general assembly, MPG President Otto Hahn announced the incorporation of the Berlin research institutes, effective 1 July 1953. Testifying to the collective pride in this achievement, the famous singer Günter Neumann was invited to perform the "song of the MPG."[13] Later, Hahn thanked Ernst Reuter in writing for his help and also took the opportunity to express his sympathy for the uprising in East Berlin on 17 June 1953.[14] To this Hahn added the grandiose fantasy that the incorporation of the DFH might be the first step toward German reunification and, thereby, initiate a return of the MPG to Berlin. More concretely came a guarantee from the MPG of the continued existence of the Berlin institutes, which could henceforth only be relocated or dissolved by mutual agreement between the city and the Society.

---

11 The Göttingen MPI for Physical Chemistry merged in 1971 with the MPI for Spectroscopy and the resulting institute then took the name of MPI for Biophysical Chemistry.

12 Bericht der Berliner Kommission aus der Niederschrift des wiss. Rates der MPG of 19 May 1953. Auszug Niederschrift der Senatssitzung der MPG, 20 May 1953, p. 11, MPGA, Abt. IA, 9/1 /3 Berliner Angelegenheiten, Korrespondenz der Geschäftsstelle Berlin, 1953–1957.

13 Marianne Reinold to Heinz Pollay, 22 July 1953. Ibid.

14 Hahn to Reuter, 25 June 1953, Ibid.

Fig. 4.6. *Max von Laue, circa 1954.*

Laue's efforts on behalf of the Institute could now be carried out within a stable organizational framework, and as a result, the staff and facilities of the Institute expanded steadily. Nonetheless, the isolation of West Berlin, a Cold-War border town surrounded by a communist state, made it challenging to attract internationally renowned scientists to the Fritz Haber Institute (FHI). As a result, new members of the scientific staff were frequently quite young and still working toward recognition when they arrived at the Institute. Some of them came directly from Berlin; some were refugees from the GDR. The territorial and political isolation of West Berlin also hindered the exchange of personnel between the Institute and other parts of the international research community, increasing the likelihood that scientists would spend their entire career in Dahlem. Furthermore, the departments during the 1950s formed a somewhat loose, heterogeneous conglomeration, which was to a significant extent the result of the frequent changes in institutional leadership between 1945 and 1959.

In this last regard, however, Laue attempted to steer against the wind and lend research at the Institute a tighter topical focus. Early in 1948, he had already drafted a working plan for the new MPI for Physical Chemistry in Göttingen, whose director was really supposed to be Bonhoeffer, but Bonhoeffer initially neglected the new institute and remained in Berlin. In response, Laue proposed changing the scientific orientation of the new institute, which had been tailored to Bonhoeffer's research interests, and using its rooms to set up an MPI for Structure Research:

> The goal is the investigation of matter through X-ray and electron diffraction, which does not exclude electron microscopy, when applied to this end. If research restrictions are removed, the important method of neutron diffraction could be added.[15]

---

15 Max v. Laue to Otto Hahn, Göttingen, 8 March 1948. MPGA, Abt. II, Rep. IA. Institutsbetreuerakten MPI für (bio)physikalische Chemie, Allgemein, Bd. 1, 1945 until 31 December 1959.

Fig. 4.7.   *The FHI in 1955, consisting of (left to right): glassblowers workshop, new X-ray building with extension for Ruska's department, factory building with X-ray annex, connecting corridor with main entrance and main building.*

Laue had in mind five departments: X-ray and electron investigation of gas molecules and liquids, crystal structure and electron distribution, Raman or absorption spectroscopy of gases and crystals, and a small department that would act as a bridge to the core concerns of classical physical chemistry. The object of investigation that united all the departments would be the microstructure of matter. Ernst Telschow, director of the MPG General Administration, wanted to have the plan reviewed, but only if Bonhoeffer truly was not coming to Göttingen. Bonhoeffer caught wind of the new plan and was disconcerted that it corresponded so little to his own research interests.[16] When Bonhoeffer decided for Göttingen over Berlin, Laue's plan became redundant; however, it was revived in Berlin as an outline for the further development of the FHI.

The charter adopted by the Institute in 1954 emphasized the enduring heritage of the Institute, which had taken shape within the KWG and would be carried on through the MPG, and it described the task of the FHI in general terms, as "research in physical, chemical and associated disciplines." No concrete directive could be gleaned from this description, which gave the director, who possessed "unified general control" of the Institute, substantial freedom in his choice of research fields. Furthermore, there was no fundamental change in the institutional hierarchy. The director of the Institute would be assisted by a deputy director, as well as a standing executive manager in charge of administrative tasks. Iwan Stranski and the physicist Dietrich Schmidt-Ott, respectively, occupied these two

---

16 K.F. Bonhoeffer to Otto Hahn, 16 March 1948 und Hahn an Bonhoeffer, 31 March 1948. Ibid.

Max von Laue (1879–1960)

Max von Laue, the son of a Prussian military official, grew up in various garrison towns of the German Empire, among them Strasbourg, where he graduated from high school (Gymnasium) in 1898 and began his study of physics at the local University. After an interlude in Göttingen he moved to Berlin, whose University would become his intellectual home for decades to come. In 1903 he graduated in Physics, as one of the few students of Max Planck; subsequently, he received an assistantship (1905-1909) and then held a professorship (1919-1943) in physics at the University. WWII brought him to Western Germany, but in 1951 he returned to Berlin, to the Dahlem KWI Institute for Physical Chemistry and Electrochemistry, which he would integrate into the Max Planck Society as the Fritz Haber Institute and direct until shortly before his death.

His most important scientific achievement, the discovery of X-ray diffraction by crystals, which made him famous and secured him the 1914 Nobel Prize in physics, he realized in Munich, while he was a Privatdozent at the Ludwig-Maximilians-Universität (1909-1912). It was there that he conceived the idea of testing whether a crystal could act upon X-rays the way a diffraction grating does upon normal light. He recruited his colleagues Walter Friedrich and Paul Knipping to implement the idea in the laboratory and, after months of intense work, they obtained, in the summer of 1912, the first X-ray diffraction images. The experiment demonstrated decisively that X-rays are short-wavelength electromagnetic waves and that crystals consist of three-dimensional lattices of atoms. The discovery became a focal point of physics discussions among experts and the public alike, and for Einstein, the much-acclaimed experiment belonged to the "most beautiful of what Physics has experienced." Although von Laue worked out a preliminary theory of X-ray diffraction which, despite its rudimentary character, remains the basis for crystallographic analysis, X-ray structure analysis would become the bailiwick of other researchers. Von Laue's interest lay mainly in questions of principle, hence his later contributions to X-ray research focused on X-ray optics and the dynamics of X-ray diffraction. This same predilection would also be reflected in his contributions to other aspects of physics, such as general relativity and superconductivity.

The discovery of X-ray diffraction boosted an already steep academic career climb for von Laue. A call to Zurich came in the very year of the discovery, and was followed by a professorship at the newly constituted University of Frankfurt in 1914. In 1919, von Laue succeeded in an "office-swap" with Max Born to land a professorship in Berlin, side by side with his mentor Max Planck. He would stay at the Berlin University until his retirement, while maintaining close connections to other institutions in the city. In 1920, he was elected Member of the Prussian Academy; in 1921 he became the

would lend colloid science a precisely-formulated, mathematical basis and, in the process, integrate the insights of statistical physics, mathematical physics and quantum mechanics. His aspirations were visionary – to turn paracrystal research into its own branch of science dealing with "intermediate states between the crystalline and the amorphous."[32] To this end, Hosemann and his colleagues studied the structure of imperfect solids, choosing example materials from a range of plastics, imperfect crystals and alloys. They relied upon physical research methods, particularly small- and wide-angle X-ray scattering studies, which complemented Laue's efforts and were in keeping with certain of the Institute's research traditions. Early on, the Indian physicist Subodh Nath Bagchi was one of the main researchers responsible for the mathematical structure of the new theory. Bagchi arrived at the Institute in 1951, in response to an invitation from Laue,[33] and departed in 1957 to take up a post as Professor of Chemical Physics at University College in Calcutta.[34] In addition to this theoretical work, the group developed new measurement techniques that enabled the spatial representation of spectroscopic data, largely thanks to the efforts of Heinz Barth.[35] The guestbook from the department shows that Hosemann's activities attracted considerable international interest, particularly from the private sector. The department also gave rise to several spin-off enterprises. During his time at the FHI, Hans Bradaczek developed an X-ray generator that was later produced by *Richard Seifert & Co*. Then in 1961, Bradaczek founded *EFG Röntgentechnik*. Similarly, Harald Warrikhoff, who had built a photoelectric radiation dosimeter during his time at the FHI, went on to establish *Röntgen-Technik Dr. Warrikhoff KG*.[36] In toto, Hosemann's department carried out an impressive range of theoretical and technical work. Nevertheless, Hosemann failed to establish a new discipline, or even a stable research school, based on his research into paracrystals. This occurred partly for scientific reasons, but was apparently also hampered by Hosemann's unwillingness to compromise and, not least of all, by the territorial and political isolation of West Berlin.

Much like Rolf Hosemann, and in spite of his international scientific reputation and his many years of service at the Institute, Kurt Molière had to wait an unusually long time before becoming a Scientific Member. Molière only received membership in 1960, thanks to an application submitted by the new director of the Institute, Rudolf Brill.[37] In the early 1950s, Molière and his department intensified research into the physical foundations of electron diffraction using a newly constructed diffraction apparatus for medium-velocity electrons. In the 1960s, they began performing diffraction experiments using slow electrons, which had to be conducted under difficult-to-maintain high-vacuum conditions but offered important insights into crystal surfaces. However, Molière's main research interests

---

32 Hosemann, *Lattice*.
33 Bagchi to Laue, 4 October 1951, MPGA, Nachlass Laue, Nr. 197.
34 Bagchi to Laue, 29 December 1959, Ibid.
35 Barth, Hosemann, *Parallelstrahlmethode*.
36 Barth, Hosemann, *Parallelstrahlmethode*.
37 Sektionsprotokoll, 9 May 1960, MPGA, CPT Akten, Niederschriften der Sitzungen.

Fig. 4.9. *Iwan Stranski (left) and Kurt Molière, 2 October 1961.*

remained applications of the dynamical theory of electron diffraction to ideal crystals, the kinematic theory of electron scattering and numerical methods for the study of the fine structure of electron reflection.

When it came to experimental techniques, Molière relied on the help of well-trained and long-serving staff members such as Günther Lehmpfuhl. Lehmpfuhl had studied physics at the Freie Universität, but he was already working at the FHI when he began writing his diploma thesis in 1953. He became a senior research assistant at the Institute in 1963. He specialized in experiments on adsorption phenomena and multi-beam electron diffraction setups. Molière's and Borrmann's departments tended to produce a limited number of relatively comprehensive publications within their respective fields of specialization. It was a well-respected approach at the time, as illustrated by the invitation Lehmpfuhl received from Albert Lloyd George Rees to spend half a year working at the laboratory of the Commonwealth Scientific and Industrial Research Organisation in Melbourne in 1965 and by Lehmpfuhl subsequently being offered a permanent position in the organization. Late in the 1960s, researchers at the Institute increasingly came to rely upon computers to generate theoretical data and models that were then compared with experimental results. One of Molière's key assistants in this line of work was Kyozaburo Kambe, who developed new theories and algorithms that were converted into computer programs.[38]

Laue not only worked to establish a new approach to research at the Institute but also went to great lengths to modernize the technical facilites of the Institute. The old equipment in the machine hall was removed between 1954 and 1956 and replaced by modern equipment from *Siemens & Halske*. As a continuation of this process, the Institute was able to afford and install between 1957 and 1958 a

---

38 Z.B. Kambe, *CellularMethod*.

Fig. 4.10. *Refitting the "Machine Hall," circa 1955.*

four-liter helium condenser built by Linde AG based on the Meißner design and a hydrogen and nitrogen liquefaction plant. Laue hoped that the ready availability of liquid helium would enable more intensive work on superconductivity.[39] He chose Gustav Klipping, a student of Stranski's trained in engineering, to oversee the liquefication plant.[40] Klipping required some further instruction to fulfill his new duties, but he was soon able to operate the plant successfully, and with the help of his wife Ingrid (née Karutz) whom he had met at the Institute, he was able to develop it into a largely independent cryogenic laboratory. On the occasion of their engagement in 1957, Laue wrote: "Shouldn't the marriages forged at the Institute also be mentioned in the Institute's annual reports?"[41] Klipping developed and patented new cooling methods and cryogenic equipment, but also conducted scientific research. Cryogenic procedures became something of a specialty of the Institute and were available to, and actively used by, all its departments. Futhermore, the cryogenic laboratory became a vital regional scientific resource since it remained the only facility of its kind in Berlin for a number of years, which enabled Klipping to develop an extensive network of contacts.[42]

Among the many department heads from the early days of the FHI, Ivan Stranski enjoyed a particularly strong scientific reputation. He was also a close friend

---

39 Zeitz, *Laue*, p. 178.
40 Klipping, *Hexamethylentetramin*.
41 Laue to Dietrich Schmidt-Ott, 16 August 1957. MPGA, Nachlass Laue, Nr. 1776.
42 Komarek, *Klipping*.

Fig. 4.11.   *The FHI helium liquefaction plant in the machine hall, 1958.*

and ally of Laue and became his proxy in matters of institute management after the FHI was incorporated into the Max Planck Society.[43] As an expert on crystal growth and crystal surfaces, Stranski fit well with Laue's notion of structure research. He was also a key figure in the post-war Berlin scientific community who had earned international acclaim. His department at the FHI, which he administered conjointly with his activities at the Technical University, investigated the crystal structures of sundry organic and inorganic compounds but with a special focus on modifications of arsenic, of which they found some new examples. Theoretical calculations together with experiments on nucleation and crystal growth, as well as phase transitions, adsorption, sublimation and solvation produced new insights into the structures of crystal surfaces and the processes through which they could change. A related topic that received a great deal of attention was triboluminescence. Stranski's department also performed X-ray structure analyses, focusing on cobalt and nickel complexes, as well as studying organic molecules. In a clear tribute to the Institute's director, they also determined the structure of the phosphate mineral Laueite.[44]

Before the age of computers, X-ray structure analysis involved a multitude of time-consuming calculations employing Fourier transforms. These calculations were among the central activities of Kurt Becker, who was installed as assistant

---

43   Sektionsprotokoll, 19 May 1953, MPGA, CPT Akten, Niederschriften der Sitzungen.
44   Plieth, Ruban, Smolczyk, *Laueit*.

Fig. 4.12. *"Mister, could you chill my beer for me?" Cartoon concerning Helium liquefaction at the FHI.*

to Stranski in 1951, shortly after completing his dissertation,[45] and promoted to senior assistant in 1955. Becker also supervised numerous dissertations and theses together with Karl Plieth, who was also a student of Stranski and worked initially as an associate professor at the Technical University, then after 1966, as a full professor of crystallography at the Freie Universität. In 1964/65, Becker was a guest lecturer at MIT in Boston, where he got the idea to start a new research program on the reaction kinetics of heterogeneous molecular sieve catalysis. When Stranski retired, early in 1967, Becker took over his research group, and the TU appointed him Adjunct Professor of Physical Chemistry in 1968.

Kurt Überreiter, who had been at the Institute since the late 1930s and who had contributed critically to the preservation of the Institute through his activities as President of the ill-fated German Research University, was appointed a Scientific Member of the Max Planck Society in 1954. Überreiter's field of research, the structure and properties of high polymers, was influenced by his work with Ernst Jenckel but also fit well with Laue's vision of the Institute. However, Überreiter oriented his research more towards chemistry, and his analytical methods were more closely related to thermodynamics and included measurements of vapor pressure and melting point depression, viscometry and dilatometry, in contrast

---

45 Becker, *Claudetit*.

to the predominantly spectrometric research conducted in the other departments. Gerhard Kanig was a close colleague of Überreiter, whose diploma research Überreiter had supervised in 1943, and who had also conducted his doctoral research at the Institute, receiving his degree in March of 1945. Kanig became Stranski's senior assistant at TU Berlin in 1958. The following year, he habilitated at the TU and accepted a position as head of a colloid chemistry laboratory at BASF.[46] Together with colleagues such as Kanig, Frithjof Asmussen and Hideto Sotobayashi, Überreiter studied the preparation, composition, phase transitions and, above all, the polymerization and solution processes of high polymers.

Not all of the research performed at the Institute fit with Laue's concept of a center for structure research. When Bonhoeffer took over the Institute in 1948, i.e. before Laue's arrival, he established a research group within his department under the guidance of his student Klaus Vetter that worked on electrochemical problems. The following year, a second electrochemistry group was added under the guidance of Georg Manecke. Vetter, who came from Berlin and completed his doctorate in 1941, reportedly as the last student of Max Bodenstein, was employed by the KWI beginning in 1946 but left later that same year to become Bonhoeffer's assistant at Berlin University. His time at the University proved decisive for the development of his scientific interests, and he returned to the Institute in 1948 as the head of a research group that focused on the classical electrochemistry of redox processes in metals and metal oxides. This included not only research on electrode kinetics but also experiments on electrochemical surface activation and deactivation, especially with iron and nickel, and experiments on metal corrosion.[47] Vetter quickly became one of Germany's leading electrochemists. In 1953, the German Bunsen Society awarded him the first of their Nernst-Haber-Bodenstein Prizes, together with Heinz Gerischer, another Bonhoeffer student who would later play an important role at the FHI. Vetter became the first full professor of physical chemistry at the Freie Universität in 1961 and moved into one wing of the former KWI for Chemistry. The professorship put him and his colleagues at the FHI into closer contact with students, making it markedly easier for them to recruit new staff. In turn, scientists from the FHI participated extensively in the development of the physical chemistry curriculum at the Berlin universities after the Second World War. In addition to his professorial duties, Vetter continued to lead his small research group at the FHI, and he was appointed an External Scientific Member of the Institute in 1966.[48]

Georg Manecke, who led the second research group under Bonhoeffer, had arrived at the KWI from Berlin University. His group specialized in ion-exchange resins, and their research focused on the synthesis of high-redox-capacity resins and the subsequent analysis of their properties and chemical structure. Manecke's group was extraordinarily accomplished at synthesis. Various carrier polymers

---

46 Springer, *Kanig*.
47 E.g. Vetter, *Korrosion*.
48 Sektionsprotokoll, 22 June 1965 and 21 June 1966, MPGA, CPT Akten, Niederschriften der Sitzungen.

Fig. 4.13. *Klaus Vetter (left) and Georg Manecke, 1957.*

were combined with functional elements, including enzymes, in such a way that reactions could take place in the stationary phase. Hence these redox resins could be used as elegant stereoselective reagents, which naturally attracted a great deal of commercial interest. For example, redox resins could be used in dialysis or in the production of hydrogen peroxide. They also played an important part in certain techniques of modern biotechnology. In 1957, Manecke left the Institute to become Professor of Macromolecular Organic Chemistry at the Freie Universität and was named an External Scientific Member of the FHI in 1963.[49] He was appointed director of the Institute of Organic Chemistry at the Freie Universität in 1964.[50] But even after his appointment at the Freie Universität, some of his employees remained at the FHI and others worked at the the Technische Universität Berlin, where Manecke was also Honorary Professor of Chemistry and Plastics Technology. The extent of his research group was probably a sign of Manecke's success in applying for external funding. Laue made no attempt to remove the research groups established under Bonhoeffer from the Institute.[51] Although both working groups were nominally part of Laue's department, they were largely autonomous, and their research interests did not overlap with the director's in any significant way.

---

49 Senatsprotokoll, 6 December 1963. MPGA, Abt. II, Rep. 1A, Senatsprotokolle.
50 Broser, *Geschichte*.
51 Zeitz, *Laue*, p. 168.

The Special Case of Ruska

In principle, Ernst Ruska's department would have complemented Laue's plan well as electron microscopy showed promise as a new method for structure research. Moreover, Ruska had won special respect for himself from Berliners when he decided to turn down attractive offers from institutions in West Germany and remain in the city, unlike the majority of renowned scientists and engineers.

In 1949, the year that his department was established at the KWI for Physical Chemistry and Electrochemistry, Ruska was also appointed Honorary Professor of Electron Optics and Microscopy at the Freie Universität and lecturer at the Technical University of Berlin, where he would become adjunct professor in 1959. At first, Ruska only worked at the institute part-time; in 1952, he asked Laue to expand his department so that he could move fully to the FHI. The move would offer Ruska the opportunity to pursue his research and development goals essentially unchecked; whereas, corporate strategists at Siemens wanted to develop mainly mass-produced instruments rather than the complex and expensive high-performance devices that Ruska had in mind.

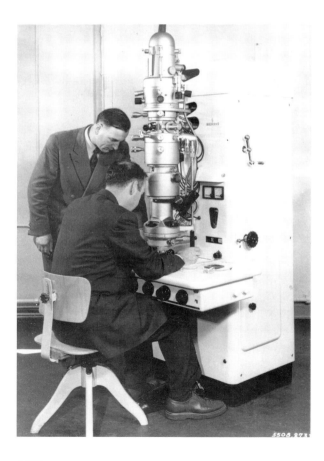

Fig. 4.14. *Ernst (left) and Helmut Ruska at the electron microscope, circa 1957.*

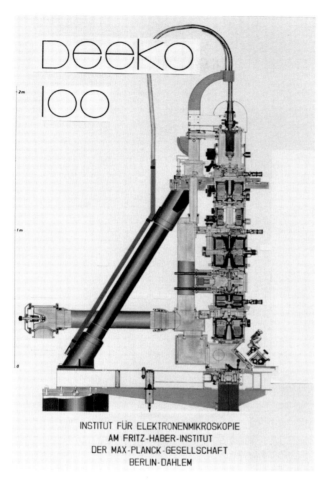

Fig. 4.16. *The DEEKO 100. Cross-section from the title page of a brochure.*

paths in connection with the construction of specific new instruments. From the 1960s onwards, the use of electronic calculators sped up markedly the necessary numerical calculations.[66]

Ruska's pet project was the development of a completely-new, short-focal-length, electromagnetic, single-field condenser-objective where the object lay at the focus of the magnetic lens. The top part of the device functioned as a condenser lens, the bottom part as an objective lens. This arrangement is now standard in electron microscopy. Both structural details and adjustable parameters were recalibrated to support this sensitive, high-performance system. An electro-static lens that functioned as a stigmator was then added to the objective to correct image asymmetries. The result was the DEEKO 100, completed in 1965, a transmission

---

66 Preisberg, *Simulation*.

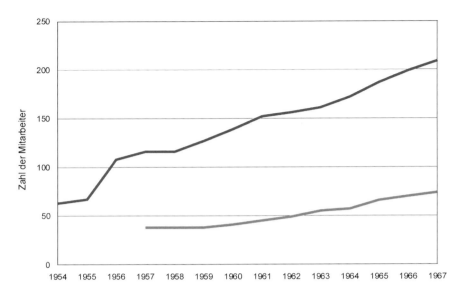

Fig. 4.17. *The number of total staff employed by the FHI (top) and the IFE (bottom). Note the sharp increase in the early MPG-era, 1955–1957. Source: Jahresberichte of the FHI and the IFE.*

electron microscope with an accelerating voltage of 100 kV that enabled magnification of 800,000 : 1. Its high performance was the result of a level of technical precision that brought with it certain problems, namely the need for highly-stable and reliable voltage sources and electronic measurement and control instruments. The considerable length of the instrument also made it susceptible to vibrations. Hence, the microscope could only achieve its highest resolution, 2.5 Ångström, during the night, when ground vibrations were at a minimum. In light of this problem, Riecke and his team conducted initial experiments on vibration-damping suspension systems. Hans Günther Heide, one of Ruska's most talented construction technicians, took a different approach; he disconnected the adjustment mechanism from the support table. Thanks to this, and other mechanical changes, the vibration sensitivity of the Elmiskop I was significantly reduced.

The technological emphasis that Ruska introduced to the Institute did not fit well with Laue's concept of structure research, nor was it in line with the Max Planck Society's self-image as a basic research institution.[67] However, thanks to his technical expertise and his international renown as a pioneer in electron microscopy – not to mention being a proud Berliner – Ruska enjoyed something of a favored status in the city and among its political elites. Discussions began with Ruska as early

---

67 It was decided in 1956 that the Institute for Scientific Instrumentation in Göttingen should be divested from the MPG, as it was not an MPI as defined by the MPG.

Fig. 4.18. *The site being prepared for the construction of the library and administration building, 10 March 1958. On the left, the main building of the FHI. On the right, the Minerva Tower and, far right, the "Radium Hut" of the former KWI for Chemistry.*

as 1953 concerning the possibility of separating his department from the FHI and upgrading it into an independent MPI for Electron Microscopy.[68]

Because Ruska took little interest in science organization and policy, he favored a compromise between maximum autonomy and minimal administrative duties. As a result, his department became the Institute for Electron Microscopy (IFE) at the Fritz Haber Institute,[69] i.e. an independent institute attached to the FHI. This introduced an additional level of administrative hierarchy at the FHI. As director of the IFE, Ruska was in many respects independent of the general director. The IFE had its own financial administration and drew up its own annual budget, which then had to be integrated into the overall budget of the FHI. This administrative splintering echoed the scientific heterogeneity of the FHI, and its effects were enhanced by the fact that IFE employees made up a considerable proportion of the FHI's total workforce. What, on the one hand, may have appeared an awkward and disjointed structure, on the other hand, helped enhance the infrastructure and prestige of the FHI. With the signing of a contract between the Max Planck Society and the Berlin Senate on 5 July 1957 that essentially resolved previous issues surrounding the ownership of KWG properties in Dahlem, Ruska and Laue were able to steer a large portion of the funds made available through the contract towards the construction of a generously-sized new building for the IFE.

---

68 Sektionsprotokoll, 19 May 1953, MPGA, CPT Akten, Niederschriften der Sitzungen.
69 MPG, *Jahrbuch 1958*, p. 59.

Fig. 4.19.   *Laying of the cornerstone for the Institute for Electron Microscopy at the FHI, 5 July 1957. From left to right: Max von Laue, Mayor Otto Suhr, Ernst Ruska (with hammer) and MPG Vice President Wilhelm Bötzkes.*

In order to create a suitable site for the new building the south-eastern border of the FHI grounds was straightened. In addition, the arcade across from the original Institute buildings and the Minerva Tower, erected during the Second World War to house new high-voltage facilities for the KWI for Chemistry, were demolished. The imposing four-story Electron Microscopy building set a new standard for the scale of MPI edifices in Berlin and also reflected efforts made by the City of Berlin to present itself as an emerging centre of scientific research, in spite of its difficult circumstances. On 5 July 1957, the laying of the cornerstone for the new building was celebrated in the presence of Berlin's political leaders and high-ranking representatives of the Max Planck Society.[70] In accord with Ruska's wishes, the building was ready to be unveiled to the public by the fall of 1958, when the shell of the new structure housed an exhibition of instruments as part of the 4th International Congress for Electron Microscopy, which was held in Berlin from September 10 to 17, 1958.

The funding package for the IFE also included allotments for a new library and administrative building as well as a large lecture hall. When the construction was complete, the FHI buildings almost completely surrounded the Haber Linden, and

---

70   Present at July event were, e.g. the governing Mayor of Berlin Otto Suhr and MPG Vice-President Wilhelm Bötzkes, both of whom died briefly thereafter. MPG, *Jahrbuch* 1958, p. 59.

Fig. 4.23. *The FHI in September 1961. Library building, IFE, X-ray annex from 1928, and factory building from 1912. On the left, the Haber Linden.*

remained in the director's department and continued to be guided by Gustav Klipping with relative autonomy. But Brill also established two new research groups. The first, working in radiography, was headed by the chemist Hans Dietrich, who had come to Laue's department from Heidelberg in 1957 and, upon arrival, had been sent to Oxford to learn X-ray crystallographic structure determination in Dorothy Hodgkin's lab. Under Brill, he and his colleagues primarily investigated the structure of catalytically significant organometallic complexes. The Dietrich group also worked with the Hahn-Meitner Institute (HMI), which had been established in 1959 in Wannsee, largely on Laue's initiative, and operated a research nuclear reactor.[80] Work began in 1960 on a neutron diffraction apparatus that would use the HMI research reactor as its source and be operated by the FHI. As part of the project, Dietrich's group developed and installed an automatic neutron diffractometer at the reactor. With the support of the HMI's department for reactor physics, a second beamline was installed for the FHI, and with the help of newly developed neutron optics, the structure of decaborane could be determined.[81]

It was not particularly beneficial to the unity of the Institute that Brill selected surface catalysis rather than structure research as the new focus of his activities. No research had been carried out on surface catalysis at the Institute since the Haber era. Nonetheless, Brill gave high priority to research on the reaction mechanism of the Haber-Bosch process over an iron catalyst. Using modern analytical

---

80 Weiss, *Großforschung*.
81 Hamilton, *Neutron Diffraction*. Brill, Dietrich, Dierks, *Bindungselektronen*.

Fig. 4.24.   *Otto Hahn in the new library of the FHI, 9 October 1963.*

methods, Brill believed he had demonstrated that the hydrogen adsorbed on the metal surface combined with the nitrogen molecules to form higher nitrogen hydrides which then broke down to form ammonia in the final phase of the reaction.[82] He also examined the influence of activators, promoters and contact poisons on the ammonia catalyst and on other catalysts.

In 1964, Brill created a second new research group, this time dedicated to field electron emission spectroscopy. It was initially headed by Werner A. Schmidt, who moved to Brill's department from the remainder of Erwin Müller's group, which had become part of Borrmann's department. In 1966, Jochen H. Block took over the group, joining what would become a 50-year tradition of field electron emission spectroscopy at the Institute, spanning from end of the 1940s into the 1990s. Brill sought to groom Block for leadership, recognizing that the latter was interested primarily in spectroscopic methods, especially field-ion mass spectroscopy, that could be used for surface analysis and, hence, might complement Brill's work on heterogeneous catalysis.

Block had completed his doctorate in 1954 under the father of German catalysis research, Georg-Maria Schwab, and remained in Munich through 1960 to complete

---

82 Brill, *Ammoniak.*

his habilitation. He worked subsequently for the European Research Association in Brussels and spent a year in the U.S. as a consultant for Union Carbide. Block and his group conducted experiments with field electron microscopes and mass spectrometers, seeking both to develop more advanced research techniques and to explain the mechanisms of surface catalytic reactions. For example, their mass spectrometer findings, together with infrared spectra, appeared to support Brill's hypothesis concerning the hydrogenation of nitrogen molecules adsorbed on iron surfaces.[83]

In 1965, Brill applied for an extension of his directorship up to his 70th birthday so that he might complete important research projects already underway. He also mentioned in his application that he would need to help Block establish himself at the Institute after his arrival on 1 January 1966 and that the question of his succession depended on Block's professional development.[84] The MPG President granted him an extension on these grounds, and Brill remained in charge of the Institute until 1969.

The orientation of the Institute toward structure research was further diminished when Ivan Stranski went emeritus in 1967. His department was formally incorporated into Brill's department, forfeiting its independent status, and was allowed to gradually fade away. This prompted a scientific reorientation of the FHI, which initially led to further fragmentation and was the cause of much concern at the Institute. In light of the complexity of the situation, the CPT Section established not just a search committee to find a successor to Brill but a so-called "core committee" to consider the future shape of the Institute. As was increasingly the case at larger MPG institutes, it was no longer possible for a single individual to speak for all the research interests represented at the FHI. The task of the core committee, therefore, was not only to find a successor to the director, but also to develop a new, overarching organizational and research plan for the Institute.

Independent of the core committee, the heads of those departments involved in structural research, under the guidance of Hosemann, composed their own proposal for the Institute. Their vision of the future development of the FHI harkened back to Laue's earlier plans and involved once again building up the Institute into a center for the use of diffraction and interference methods in atomic and molecular structure research.[85] In 1967, their concept paper was passed on to MPG President Adolf Butenandt, who had taken office in 1960. It was also pointed out to him that the proposal it laid out was in step with new developments in the international research community; materials science was enjoying strong backing, particularly in the United States.[86] At least in the fields of electron and X-ray

---

83 Schmidt, *Massenspektrometrie*. Brill, Jiru, Schulz, *Infrarotspektren*.

84 Brill to MPG President Adolf Butenandt, 15 June 1965. MPGA, Abt. II, Rep. 1A –IA5/–, Handakten Gentner, Zukunft des FHI – Nachfolge von Prof. Brill, Korrespondenz/Gutachten.

85 Borrmann and Hosemann to Sektions- und Kommissionsvorsitzenden Wolfgang Gentner, 17 May 1968. Ibid., "unprocessed".

86 Ibid., Anlage: R. Hosemann, Zur Zukunft des Fritz-Haber-Institutes (Ausarbeitung einer Aktennotiz from 31 March 1967).

Fig. 4.25. *Rolf Hosemann, recumbent on the right, with his department, 1968.*

diffraction, the FHI could become a counterpart to the Laue-Langevin Institute in Grenoble, where structural research focused primarily on the use of neutrons. In this context, Brill was accused of having failed to build on the foundations laid by Laue, as a result of which "in the following years, the institute [had] lost uniformity and focus."

Hosemann and Borrmann also brought this new version of an old plan to the attention of the core committee. The members of the committee met at the FHI for the first time on 15 January 1968 to look around the Institute[87] and to gather the views of the scientific members, but not the other coworkers of the scientific staff, on the future of the Institute.[88] Borrmann, Hosemann, Molière and Über-reiter described their areas of research and presented their outlines for the future of the FHI, which were largely consistent with each other and similar to the ideas presented in the 1967 concept paper. They also called for the introduction of collaborative administration at the Institute, likely in large part because they wanted their individual departments to be as independent as possible. During later internal committee discussions, Brill expressed the view that the lines of research pursued by Borrmann and Überreiter were no longer opening new vistas for scientific research and should be abandoned with the retirement of these department heads. He advocated, instead, an intensive focus on pioneering new areas of research in

---

87 Sektionsprotokoll, 23 February 1968, MPGA, CPT Akten, Niederschriften der Sitzungen.
88 Niederschrift über die Sitzung der Kommission „Zukunft des Fritz-Haber-Instituts, Berlin – Nach-folge von Herrn Prof. Dr. R. Brill", 15 January 1968, MPGA Abt. II, Rep. 1A –IA5/-, Handakten Gentner, Zukunft des FHI – Nachfolge von Prof. Brill, Protokolle/Unterlagen zu Kommission-ssitzungen/Einladungen und Anwesenheitslisten.

catalysis, the kinetics of fast surface reactions and field-emission mass spectrometry, which was an implicit endorsement of his protégé Block. Finally, he declared himself in favor of the appointment of Hans-Dieter Beckey, who, together with Block, would represent the forward-looking faction at the Institute. This roughly corresponded with the view of the committee, underscored by the fact that the expert opinions they received on the matter were consistently positive.[89] More concretely, the core committee proposed a tripartite Institute with a new section to be established under Beckey and Block, the IFE under Ruska as the second section, and a third section encompassing the departments of Borrmann, Hosemann and Überreiter, which were no longer considered productive. The position of Molière's department remained unclear; it was proposed that he choose which section to join. Only at the end of the meeting was Ruska asked for his opinion; he approved of the plan.

The restructuring plan placed at risk the positions of certain long-time employees of the Institute who lacked permanent work contracts. Their plight quickly caught the attention of the department heads, who held a meeting to discuss the situation as early as 1966.[90] Brill encouraged these employees to complete their habilitations so that they would be able to apply for professorships or other positions commensurate with their academic qualifications; of course, this would also create space for new employees at the Institute. Meanwhile, the department heads and the affected employees worked toward establishing permanent positions at the Institute.

The first candidate proposed for the directorship was the physical chemistry professor in Bonn, Hans-Dieter Beckey. Beckey had achieved prominence through the development of new and innovative methods in mass spectrometry and fashioned his Bonn institute into a leading center for research in the field with a promising future; as a result, he could be quite confident of his future in Bonn. Nevertheless, he showed great interest in the offer from the Max Planck Society and the prospect of leading the reorganization of the FHI. As a representative of a new generation of scientists, however, he made it clear that he had no interest in burdening himself with the extensive administrative duties associated with sole direction of such a large institute. This was an easy demand to accommodate as it paved the way for the implementation of a collaborative administration. Completely in agreement with Brill, Beckey proposed the physics and physical chemistry of surfaces as the future focus of research at the Institute.[91] He proposed Erwin Müller as his colleague on the board of directors. As mentioned earlier, Müller had worked at the Institute at the beginning of the 1950s and had maintained contact with his former colleagues at the FHI after he moved to the United States, in part through regular trips to Berlin. His appointment as

89 Z.B. Robert Haul to Wolfgang Gentner, 29 October 1968, MPGA Abt. II, Rep. 1A –IA5/-, Handakten Gentner, Zukunft des FHI – Nachfolge von Prof. Brill, Korrespondenz/Gutachten.
90 Molière to Brill, 7 February 1967, Ibid., "unprocessed".
91 Molière to Brill, 7 February 1967, Ibid., "unprocessed".

an External Member of the Institute in 1957 was hardly accidental.[92] In addition, Müller had criticized Borrmann for not fully exhausting the potential of the former electron microscopy department.[93] In response to the proposal, however, Müller quickly made it clear that he would rather stay in the U.S. and added that he considered Beckey and Gerischer to be very good candidates for the directorship; although, he doubted that either would be willing to move to Dahlem. His American perspective also allowed him to recognize an almost insurmountable barrier that stood in the way of a fundamental reorientation of the Institute: "Labor law restrictions are likely to be a serious problem for any new director."[94]

At the end of 1968, the CPT Section made a concrete proposal to divide the FHI into sections for surface chemistry and physics (to which Molière now belonged), electron microscopy and structure research, and to place the entire Institute under a collaborative administration. However, the proposed organizational changes, which were manifest in the new charter of the Institute, were only to be implemented with the cooperation of the new directors. In 1968, permanent posts for scientists at the FHI were distributed as follows (IFE not included):[95]

Table 4.1. *Staff positions at the FHI.*

Structure Research		Surfaces, Electrochemistry		Services	
Department Becker (formerly Stranski)	17	Department Brill	25	Cryogenic lab, Klipping	20
Department Borrmann	13	Department Molière	13	Workshops	49
Department Hosemann	18	Work group Manecke	7	Administration	6
Department Überreiter	20	Work group Vetter	4	Computing centre	4
				Computing centre	2
Total:	68		49		81

As deliberations concerning the future of the Institute came to a climax, an event took place that served as a reminder of the long and protean history of the Institute. On 9 December 1968, the Institute celebrated the centenary of Fritz Haber's birth with a commemorative ceremony held in its new lecture hall.[96] The affair was attended not only by leading figures in the Max Planck Society and

---

92 Molière to Brill, 7 February 1967, Ibid., "unprocessed".
93 Müller to W. Gentner, 15 February 1967 (eigentlich 1968). MPGA Abt. II, Rep. 1A –IA5/-, Handakten Gentner, Zukunft des FHI – Nachfolge von Prof. Brill, Korrespondenz/Gutachten.
94 Ibid.
95 Fritz-Haber-Institut. Personal nach Stand von 14.5. 1968 mit Zahl der unbesetzten Stellen. Ibid., Protokolle/Unterlagen zu Kommissionssitzungen/Einladungen und Anwesenheitslisten.
96 MPG, *MiMax* 6/1969, p. 326–352.

Fig. 4.26. *100th birthday of Fritz Haber, 9 December 1968. Left to right: Adolf Butenandt, Michael Polanyi and Director Rudolf Brill.*

other distinguished scientists but also by numerous representatives of the political and industrial elite. Among the guests of honor was Michael Polanyi, who had resumed contact with the Institute in the post-war era and, in 1953, expressed to Laue how pleased he was with the renaming of the Institute.[97]

Even though Laue had already been dead for eight years, the commemorative event marks something of a symbolic end to the Laue era at the Institute. Brill had asked for an early retirement, to take effect in the spring of 1969, and had at the same time requested the appointment of Block as Scientific Member of the Institute. Beckey was consulted about the appointment since he was still being considered for the directorship at the time.[98] In spring of 1969, however, Beckey decided to remain in Bonn rather than move to Berlin. Heinz Gerischer was quickly selected as an alternative candidate and, after comprehensive talks were held and the necessary opinions sought, was eventually made the official candidate for Brill's successor.[99] Most likely, it was not only Gerischer's scientific qualifications that turned opinion in his favor but also his support of the prevailing plan to develop the Institute into a center for surface physics and chemistry. Gerischer also declared himself willing to hand over management of the Institute to a board of directors after a transitional period of unspecified duration. Moreover, his proposals for the scientific orientation of the Institute meshed well with the MPG's plans

---

97 Polanyi to Laue, 26 November 1952, MPGA, Nachlass Laue, Nr. 1554.
98 Sektionsprotokoll, 20 February 1969, MPGA, CPT Akten, Niederschriften der Sitzungen.
99 Gentner an Gerischer, 10 March 1969 und Gerischer an Gentner, 21 May 1969, MPGA Abt. II, Rep. 1A –IA5/-, Handakten Gentner, Zukunft des FHI – Nachfolge von Prof. Brill, "unprocessed".

Fig. 4.27. *Installation of Heinz Gerischer as director, 9 December 1968. From left to right, first row: A. Butenandt (with chain of office), H. Gerischer, R. Gerischer; second row: K.H. Herrmann K. Überreiter, I. Stranski. Fourth row on the right: R. Hosemann.*

to build up research into solids, including the creation of an Institute for Solid-State Physics in Stuttgart. These plans were formulated in reaction to German shortcomings in solid-state research, particularly in the area of semiconductors and microelectronics, that had been pointed out by various research policy bodies. Both the Stuttgart institute, which opened in 1972, and the new orientation of the FHI were intended to compensate as quickly as possible for these shortfalls.

It was against this backdrop that the CPT Section, in the summer of 1969, swiftly came to the decision to recommend the appointment of Heinz Gerischer as director of the Fritz Haber Institute to the Senate of the Max Planck Society. The appointment of Block as a scientific member was proposed at the same meeting and made dependent upon Gerischer accepting the position.[100] This deal highlights how decisions regarding personnel (*Personalprinzip*) could be closely linked to decisions regarding scientific orientation. In this case, the decision concerning the orientation of the Institute (*Sachprinzip*) had already been made before the talks with Gerischer were held, with Brill playing a key role.

The ensuing steps followed in quick succession. Brill's tenure as director ended on 31 March 1969, and Ernst Ruska took provisional control of the Institute.[101]

---

100 Sektionsprotokoll, 11 June 1969, MPGA, CPT Akten, Niederschriften der Sitzungen.
101 MPG, *MiMax* 1/1969, p. 52 and MPG, *MiMax* 6/1969, p. 361. Butenandt to Brill, 28 March 1969. President Butenandt to Ernst Ruska, 28 March 1969, MPGA Abt. II, Rep. 1A –IA5/-, Handakten Gentner, Zukunft des FHI – Nachfolge von Prof. Brill, Korrespondenz/Gutachten.

laboratory finally followed him to the FU Physics Department. Nevertheless, supply contracts with the FHI, which contributed to the funding of the facility, and with the Hahn-Meitner Institute and BESSY would be arranged.

Gerischer carried on with the electrochemical research program previously outlined, investigating electrode processes as well as photoexcitation, and set his own research group on a course aimed at developing new fundamental insights. He oversaw ongoing research on the kinetics of rapid reactions. In addition, a new temperature jump method using an iodine laser was developed which offered a wider range of application than the standard technique at the time, extending down into the sub-nanosecond range. This method was used, for example, to study the dissociation of water, demonstrating for the first time its direct photolysis.[5] Another object of investigation was the temperature-dependent dynamics of the structure of lipid double-layers, the basic building blocks of all biological membranes.[6]

Stimulated by the development of metal-oxide-semiconductor (MOS) transistors, in 1957 Gerischer began to investigate charge transfer processes in semiconductor electrodes, a topic of research upon which he came to focus more intently in Munich. By the beginning of the 1960s, he had already characterized electron transfers between metal or semiconductor electrodes and electrolytes as tunneling processes taking place through the electrical double layer.[7] This groundbreaking research was of great import for photochemistry and photovoltaics. In the same vein, Bruno Pettinger[8] and others at the FHI carried out photoexcitation studies on redox systems in solution in contact with the very-thin top layers of II–VI group semiconductor or metal electrodes; Pettinger now heads a research group in the Department of Physical Chemistry. Related research was also done on the then largely unknown electrochemical behavior of sulfide based semiconductors, as a result of which Helmut Tributsch was able to explain the mechanism by which bacteria oxidize sulfide ores.[9]

Classical electrochemical methods had the advantage that they could be applied in situ to surfaces in electrochemical cells; whereas, the low permeability of electrolytes starkly limited the use of spectroscopic methods. Hence attempts were made to investigate electrode surfaces outside the cell, in ultra-high vacuum. It was shown that surface behavior was stable under favorable conditions even in high-vacuum. This paved the way for new investigations, such as the surface-specific and angle-dependent spectroscopic studies of adsorption and oxidation states of metal and semiconductor crystals, e.g. GaAs, conducted by Karl Jacobi and Wolfgang Ranke.[10] Closely related to the investigation of metal and

---

5 Frisch, Goodall, Greenhow, Holzwarth, Knight, *Single-Photon*.

6 Eck, Genz, Holzwarth, *Iodine Laser*.

7 Gerischer, *Halbleiter I*. Gerischer, *Halbleiter II*. Gerischer, *Halbleiter III*.

8 Pettinger, *Tunnelprozesse*.

9 Tributsch, *Desintegration*.

10 Jacobi, Ranke, *GaAs Surfaces*.

Heinz Gerischer (1919 – 1994)

Heinz Gerischer was born in the "Luther city," Wittenberg, where he grew up and went to the Melanchthon-Gymnasium. After graduation, he started his study of chemistry at the University of Leipzig, which he had to interrupt after the outbreak of WWII. In 1941, however, he was able to resume his studies and obtained his diploma in 1944. As a so-called "half-Jew," he could only continue his university studies because his mentor, Karl Friedrich Bonhoeffer – along with the physicists Friedrich Hund and Werner Heisenberg as well as one member of the administrative staff – concealed this fact for him. Thus, in violation of the law of the land, he started his graduate studies as Bonhoeffer's private research assistant. In 1944-45, he was enlisted by the Todt Organization as a forced laborer. In the Fall of 1945, he returned to Bonhoeffer's laboratory as his assistant at the Institute for Physical Chemistry of the University of Leipzig and graduated in 1946 with a thesis on oscillating reactions on electrode surfaces. In the same year, he followed his mentor to Berlin as his assistant at the tradition-rich Institute for Physical Chemistry of the Berlin University. In 1949, Gerischer moved, along with Bonhoeffer, to the newly founded MPI for Physical Chemistry in Göttingen, where he remained as a staff scientist until 1953. Because of clashes between Göttingen University and the MPI about habilitation procedures, Gerischer accepted a position as department head at the MPI for Metal Research in Stuttgart and received his habilitation in 1955 at the TH Stuttgart. In 1960 he became a Scientific Member of the MPI but, in 1962, switched to the Institute for Physical Chemistry of the TH Munich as an *extraordinarius* professor; he was promoted to a full professorship in 1964. In 1969 he followed a call from the MPG to become the director of the Fritz Haber Institute in Berlin, a position he held until his retirement in 1987. He taught at both West-Berlin universities – the Technische and the Freie Universität – and held a number of guest professorships. In 1971/72 he served as president of the Bunsen Society.

During his time in Stuttgart, Gerischer developed the galvanostatic double-pulse technique, which made it possible to determine the kinetics of charge transfer independently of the diffusion and concentration polarization of the electrode. In the following years he applied a wide variety of experimental techniques to problems related to liquid-solid interfaces with a predilection for those relevant to catalysis. One of the leading electrochemists of his generation, Gerischer died in Berlin at age 75.

Fig. 5.1. *From left to right: Jochen H. Block, Heinz Gerischer, Erwin W. Müller, 1976.*

semiconductor electrodes were experiments on photoreactions at interfaces. To this end, department members analyzed photoelectron emissions from metals and semiconductors in electrolyte environments and observed direction-dependent yields that allowed them to draw conclusions concerning the behavior of excited electrons and electron holes.[11] Gerischer, together with Karl Doblhofer, also investigated and explained the mechanism of electrochemiluminescence at electrodes in nonaqueous electrolytes containing alkali and alkaline earth cations.

When the oil crisis hit in the 1970s, Gerischer's work on electrochemical solar cells became of great practical interest. With colleagues such as Helmut Tributsch and Jürgen Gobrecht, Gerischer surveyed and analyzed the potential of this means of producing energy and built strikingly effective trial "wet" solar cells. These showed that it was possible to couple light absorption to a redox reaction and that the energy produced could be stored chemically with minimal losses, e.g. as hydrogen.[12] Tributsch had done his doctorate at TH Munich and, after a research hiatus in the U.S., became a member of the scientific staff of the Department of Physical Chemistry in 1978. In 1982, he became Professor for Physical Chemistry at the Freie Universität. His interest in solar arrays and solar energy mirrored Gerischer's work in these fields. The main problem these coworkers encountered

---

11 Sass, *Photoemission*.
12 Gerischer, Gobrecht, Kautek, *Semiconducting Materials*.

in developing electrochemical cells was the poor corrosion resistance of the electrodes. Department members performed tests on various semiconductors, from which they obtained interesting results on corrosion and on the electrical properties of crystals as well as the kinetics of redox reactions at phase boundaries. The breadth of Gerischer's vision of his field was apparent in his research on the conversion of chemically stored energy using fuel cells. At the time, fuel cells employed exorbitantly expensive noble metal catalysts. Gerischer's coworkers discovered alternative organic catalysts, but because of their low conductivity these catalysts had to be stably arranged in thin layers on metal electrodes. The electrochemical analysis of polymer coatings on metal electrodes would come to be the main field of research of Karl Doblhofer, who later became head of a research group and remained at the FHI until his retirement in 2001.[13]

Work on charge transfer at boundary layers also included the study of charge injection into doped molecular crystals of condensed aromatics, such as naphthalene or perylene, for example using dye molecules excited by laser pulses. These studies demonstrated that the adsorbed dye molecules, in conjunction with an energy transfer system, could generate electric current. Gerischer investigated the mechanism and kinetics of the main steps in the attendant reactions, which were similar to those involved in a wide array of processes from photosynthesis to photographic procedures. It was hardly surprising that some of Gerischer's colleagues such as Tributsch; Frank Willig, later group leader at the Hahn Meitner Institute, and Laurence Peter, later Professor for Physical Chemistry at University of Bath, became renowned experts on solar cells, photoelectrochemistry and electrochemical energy transfer.

In light of the new overall concept for the Institute, Gerischer also built up research on interfaces and adsorption processes relevant to heterogeneous catalysis. With new spectroscopic methods, it appeared possible to analyze the local structure and organization of surface adsorbates. In light of this, department members employed techniques such as photoelectron spectroscopy, high-resolution electron loss spectroscopy (HRELS), Auger electron spectroscopy and surface enhanced Raman spectroscopy (SERS) to garner information regarding the optical, vibrational and electronic properties of interfaces. SERS in particular employed a phenomenon first observed in 1974 in specially prepared surfaces of certain metals. Pettinger worked heavily with this method and would become a specialist in SERS research.

Certain techniques, when used under ultra-high vacuum (UHV) conditions, could yield quite detailed topographical data; these techniques included low-energy electron diffraction (LEED) and reflected high-energy electron diffraction (RHEED) as well as photoemission spectroscopy using synchrotron radiation. Gerischer's colleagues employed such techniques to analyze, for example, metal single-crystal electrodes in different crystal orientations and the characteristic reflections of epitaxial metal monolayers. Among the department members who

---

13 E.g. Doblhofer, *Membrane-Type Coatings*.

worked regularly with these methods were Frank Forstmann and Dieter M. Kolb. Forstmann became an instructor at the Institute for Theoretical Physics at the Freie Universität Berlin in 1974 and later a professor there, while Kolb, who had completed his doctorate under Gerischer, became a scientific coworker in Gerischer's department in 1971 and later a research group leader. The FU Chemistry Faculty named Kolb adjunct professor in 1984, and in 1990, he became a professor and head of the Electrochemistry Department at Ulm University. In collaboration with Wilfried Schulze, who had done his doctoral research with Klipping and thereafter moved to the Department of Physical Chemistry, Forstmann and Kolb promoted the use of matrix isolation spectroscopy. This technique involved condensing atoms, molecules or their aggregates together with an inert gas, thus preserving the structures while simultaneously isolating them for spectroscopic purposes.[14] The behavior of electrons in microclusters showed particular promise as a window onto catalytic activity because corresponding theoretical models were manageable and largely calculable.

Measurements taken on adsorbate coated electrodes led to the identification of superlattices, which allowed researchers to infer the geometrical position of chemisorbed molecules. Active in this field were Karsten Horn, Alexander Bradshaw (see below) and Harm Hinrich Rotermund, who would accept an offer to become Professor for Physics at Dalhousie University in Halifax in 2003. Ultimately, through this line of research, the binding sites, degree of dissociation and spatial position, as well as the lateral movement and reorientation of molecules such as CO, O_2 and N_2 adsorbed on metal or semiconductor surfaces could be explained.

As shown by the sundry endeavors just described, Gerischer pursued electrochemical research with an uncommonly diverse range of methods. He thereby shaped his department into a hub for catalysis research true to the new orientation of the Institute. In spite of health problems, Gerischer would remain affiliated with the FHI after going emeritus in 1987.

Gerischer's colleague, Jochen H. Block, who joined Gerischer in promoting surface science and catalysis at the Institute, led an independent group of researchers specialized in his own fields of interest. He was able to form the core of what would become the Department for Surface Reactions from the resources available in the research group he had previously organized under Brill. After an initial delay, Kurt Becker and his research group also joined Block. Becker persisted in his research into heterogeneous catalysis with catalysts such as zeolites. One prominent application of these aluminosilicates was as catalysts in the petrochemical industry. Studies at the Institute concentrated primarily on their structure, stability and reactivity. However, in addition to seeking a better understanding of mechanisms of catalysis, members of the Institute also carried out experiments on reaction kinetics and catalyst poisoning. Becker's group found that the limits on the lifespan of petrochemical zeolite catalysts were set by self-poisoning

---

14 Gerischer, Kolb, Schulze, *Optical Absorption*.

Fig. 5.2. *Jochen H. Block (left) and Kurt Becker at the retirement ceremony for Becker on 25 April 1986.*

with polymerized olefins, a product of the reactions they catalyzed.[15] Hellmut G. Karge, who retired in 1996, was one of the exceptionally productive members of this group. He modified zeolites and other catalysts using ion exchange methods and thereby improved their longevity and selectivity.

The main focus of Block's interests was the behavior of surfaces in strong electric fields, which he explored using field emission phenomena, especially field-ion microscopy and field-ion mass spectrometry. Field desorption permitted inferences regarding the electronic properties of surfaces and surface adsorbates, and the atomic scale resolution of the technique allowed very precise local analysis of crystallographically well-defined surfaces. However, it required that the substrate be manufacturable in the appropriate form, i.e. thin, sharp needles. Also, since the photoexcitation of field-ion formation using light, synchrotron radiation or laser pulses (photofield emission) evinced no penetrating power, it was treated as

---

15 Karge, Ladebeck, *Mordenite*.

a technique that acted only at the surface. Institute members used this technique to study samples of tungsten, silver and aluminum hydroxide, upon which H_2, O_2, H_2S or ethylene had been adsorbed, across a range of temperatures.[16] Later experiments also examined superconductors. One of Block's abiding collaborators in these studies was Wolfgang Drachsel, who later led a research group in Hans-Joachim Freund's Department and was active at the Institute until 2004.

Another central and persistent focus of research in Block's department was heterogeneous catalysis, a field in which Block showed an early interest as cofounder and first chairman of the DECHEMA section for catalysis. As a test of the applicability of field emission methods in catalysis research, the Block group first examined simple systems like the adsorption of noble gases and the chemisorption of CO, both of which manifested significant deviations while in the apparatus from their reactivity outside an electric field. Theoretical work based on models from "high-field chemistry" and completed, in part, through a close collaboration with Hans Jürgen Kreuzer, backed up these experiments. Kreuzer was Professor for Theoretical Physics at Dalhousie University in Halifax and was named External Scientific Member of the FHI in 1998. In later experiments, the surface specific adsorption and chemisorption of various small molecules on metal surfaces were explored, phenomena that were of decisive import for heterogeneous catalysis. In order to study these systems, new methods were needed that increased the sensitivity of the apparatus. Hence, the field-ion microscope was equipped with a kind of atom probe and developed into a field-ion energy spectrometer that could generate data relating to the position and the energy of individual surface atoms at defined points on the microscope tip. One model system for induced field desorption studies was the formation of singly- and multiply-charged hydrogen ions. With linear H_3^+ ions in strong fields it was shown that the H_{3ad} species was positioned upright and linearly against the probe surface.[17] Among the coworkers who worked extensively with Block on the kinetics of reactions on metal surfaces was Norbert Kruse, who had already been in Berlin and affiliated with the FHI almost 10 years in 1977, and who is now Professor for Chemical Physics at Université Libre in Brussels.

Block placed great weight on modern technical facilities, and in the Department for Surface Reactions several instrumental methods were advanced, primarily ones related to electron field emission or ion emission. Pulse methods were developed that allowed for time-of-flight mass spectroscopy,[18] including a variant of the technique in which a laser stimulated photoemitter replaced the high voltage pulse generator for exact mass determination;[19] for the necessary UV radiation department members relied on HASYLAB at DESY and on BESSY (see below). Through the combination of pulsed desorption with time-of-flight mass spectroscopy and digitally processed image displays it was possible to determine

16 Bozdech, Ernst, Melmed, *Field Ion*.
17 Block, Bozdech, Ernst, Kato, *Formation*.
18 Abend, Block, Cocke, *Time-of-Flight*.
19 Song, *photonenstimulierte Felddesorption*.

Fig. 5.3. *Jochen H. Block, 1979.*

exactly the formation sites of various molecules. The growing import of computer technology in instrumental development can be gleaned from the fact that in 1988/89 members of the Department for Surface Reactions oversaw eight diploma theses on computers and computer networking.

In connection with research into the kinetics of the reaction between CO and O_2 on platinum surfaces a direct connection developed between the departments of Block and Gerhard Ertl. The coupled behavior of two platinum comparison crystals linked only through the gas phase was studied and showed that local oscillations in the reaction rate (see below) were, in part, synchronized through the gas phase.[20] One of the applications of the "PEEMchen," which also embodied ties to the Department for Electron Microscopy (see below), was the analysis of such phenomena. One key collaborator in work on fluctuating reactions on surfaces was Klaus Christmann, a professor at the FU Berlin and former student of Ertl. X-ray photoemission spectroscopy (XPS) was also applied to the study of surface adsorption and desorption kinetics. With this technique Institute members could, for example, attain new results pertaining to the dissociative chemisorption of nitrogen on a Fe(111) surface at the high temperatures appropriate to the Haber-Bosch process.[21] A later, more-advanced X-ray photoemission spectrometer is still operated by the Department for Inorganic Chemistry (cf. Chapter 6). In addition, in collaboration with IBM in Zurich, Institute members began experiments in 1987 intended to asses the potential of scanning tunneling microscopes, first developed six years earlier, for use in surface science research.[22]

The early death of Jochen H. Block, in the summer of 1995, leaves open the question of whether the initiatives just discussed could have launched enduring new lines of research at the FHI. On the other hand, the specialized professional

---

20 Block, Christmann, Ehsasi, Frank, *Oszillatorn*.
21 Golze, Grunze, Hirschwald, Polak, *XPS-Study*.
22 E.g. Gimzewski, Sass, Schlittler, Schott, *Scanning Tunneling*.

organizations with which he was affiliated left little question as to the high scientific regard that he enjoyed. Since 1997, the DECHEMA catalysis section, later the German Society for Catalysis, has awarded annually the Jochen Block Prize to an exceptional young scientist active in the field of catalysis research; in 1998, the prize was even awarded to FHI coworker Werner Weiß.

In the 1970s, in addition to the research groups under the auspices of Block and Gerischer, the FHI also included departments and groups that, in some facets of their activities, manifested much older research traditions. In Kurt Molière's department, which was certainly to be counted among these, work on elastic and inelastic diffraction of fast (HEED) and slow (LEED) electrons carried on a quite long-standing tradition. Molière had been active and well-reputed in this field since the 1930s. Since LEED was also superbly suited to the investigation of surface energies, this line of research also fit, in principle, with the new research program in the (sub-)Institute for Physical Chemistry. Correspondingly, the strong point of research in the department would be the physics and chemistry of solid surfaces, especially studies of the structure of crystal surfaces and adsorption layers, to which end photoemssion spectroscopy techniques would also be deployed. In addition, angle resolved methods helped in determining spatial relations at surfaces and during surface reactions.[23] In specific cases, this work overlapped thematically with that of Gerischer and his coworkers, resulting in discrete collaborations. In connection with HEED and electron optics there was also some cooperation with the (sub-)Institute for Electron Microscopy.[24] All the while, long-time coworkers such as Günther Lehmpfuhl and Kyozaburo Kambe furthered the well-established lines of research surrounding the study of crystal structures using high energy electron diffraction phenomena. In connection with this work, analyses were carried out on theoretically complex arrangements, chiefly with the aid of Bloch waves and electron density distribution calculations.[25] True to the times, computers played an ever greater role in this work. These collaborations and technical offshoots would ease the incorporation of Molière's former coworkers into other divisions of the FHI after the dissolution of the Department for Electron Diffraction. For example, Kyozaburo Kambe, dissertation advisor to Matthias Scheffler, moved to the newly founded Theory Department in 1988 (cf. Chapter 6).

Under Kurt Überreiter further research was carried out on the thermodynamic properties of polymers and their solutions. In addition, Überreiter oversaw research into topics such as crystallization delaying and accelerating structural elements and the relationship between molecular structure and macroscopic properties. This included the first measurement in polymer solutions of effects from configuration on surface tension.[26] Members of the department also built a wide array of instruments, such as high pressure dilatometers,[27] differential refractometers

---

23 Forstmann, Kambe, Scheffler, *Angle Resolved Photoemission*.
24 Fujimoto, Kambe, Lehmpfuhl, Uchida, *Dunkelfeldtechnik*.
25 E.g. Fujimoto, Kambe, Lehmpfuhl, *Electron Channeling*.
26 Überreiter, Yamaura, *Surface Tension*.
27 Karl, *Hochdruckdilatometer*.

Fig. 5.4.   *From left to right: Matthias Scheffler, Karl Doblhofer, Edith and Kurt Molière and Elmar Zeitler, 1992.*

and rotation ebulliometers; while new, computerized Monte-Carlo calculations enabled theoretical contributions to our understanding of the internal structure of macromolecules. Among the contributors to these calculations were Hideto Soto-bayashi, who was already at the Institute in 1957 and stayed until his retirement in 1994, thereafter remaining a resident of Berlin. As a young student, Sotobayashi survived the detonation of the atomic bomb over Hiroshima, on 6 August 1945, and in recent years he has begun using powerful, personal accounts to warn of the unforeseen consequences of an imprudent exploitation of our knowledge of atomic physics.

Gerhard Borrmann, who already preferred to pursue his activities at the Institute unobtrusively in 1951, retired early in 1970. His department continued to exist, at least formally, under the guidance of Gerhard Hildebrandt, who pressed on with the investigation of X-ray fine structure. Michael Drechsler and his group also belonged to this remainder of the Borrmann Department, which would be incorporated into the Hosemann Group in 1974. There Hildebrandt would initiate work on X-ray topography using both conventional and synchrotron radiation sources. At first his group carried out their experiments using the DORIS storage ring at DESY in Hamburg, but after DESY opened HASYLAB in 1981, they were granted permission to use a work station there.

The core of Hosemann's Department, which moved from the Institute's old campus into the back building of the former KWI for Fiber Chemistry in 1971, persisted in its pursuit of paracrystal research. They deployed new theory-based

Fig. 5.5. *Rolf Hosemann (left) and Gerhard Hildebrandt, 1987.*

rules, methods of analysis and computer aided renditions to refine their structural analyses of synthetic polymers, biopolymers, catalysts and melts. They intended, thereby, to offer further proof of the prevalence in nature of the microparacrystalline and point to advantages that the new concept held over various established doctrines. All the same, Hosemann's campaign for broader support of paracrystal studies remained unsuccessful. After disbanding the Paracrystals Department in 1980, Hosemann was nevertheless able to establish an Institute for Paracrystal Research at the nearby Federal Office for Materials Research (*Bundesanstalt für Materialforschung und -prüfung*, BAM) and carry forward his investigations until forced by his health to step down in 1987.

The Institute for Electron Microscopy (IFE) – between Science and Technology

The unavoidable retirement of Ernst Ruska was set to occur at the end of 1974. At the same time, pressure for the timely reorientation of the FHI was increasing thanks to the activities of the new MPG President, Reimar Lüst, who pursued with marked vigor the reforms that Butenandt had promoted only cautiously. This elicited noticeable concern from many older, more-conservative members of the MPG. At the top of the new agenda, alongside modernizing the research orientation and scientific administration of the institutes, stood promoting the involvement of scientific staff in decision making processes. As a result, scientific coworkers at the FHI were for the first time to be kept officially informed of decisions under consideration that might affect the Institute, and Hellmut Karge, who had arrived at the FHI as one of Stranski's doctoral students, was elected the scientific coworkers' representative to the CPT Section; thereafter, he would be elected their representative to the MPG Senate as well. These changes lent new impetus to the plans for reorganization under Gerischer. Lüst also stood by the provision of the Society's charter that barred raising the retirement age for scientific members, and thus assured the pending generation change and concomitant institutional restructuring at the FHI and beyond. In the case of the FHI, this stance did much to curb the persistent resistance of Rolf Hosemann to the new orientation of the Institute, which he continued to express in forms ranging from counterproposals to press releases.

There were similar differences of opinion concerning provisions for a successor to Ruska. As was so often the case, a personnel decision was linked with a question of subject matter, and fundamental changes in the direction of the IFE were deliberated. Ruska's coworkers were schooled in a highly-specialized manner of work that conformed more closely to industrial practices than those of a scientific research institute. Nevertheless, thanks to the renown of Ruska, his ability to fabricate groundbreaking instruments and the accumulated know-how of his expert staff, the IFE enjoyed a singular international reputation, even if many scientists felt its technical orientation somewhat excessive. The CPT Section established two separate committees, one to find a successor for Ruska and the other to devise an outline for the future development of the FHI; however, both committees worked hand in hand. The majority of experts surveyed were of the opinion that electron microscopy at the FHI should be more closely tied to ongoing experimental research. Whereas Ruska, along with the majority of his coworkers, did not want to see a radical change in the orientation of the IFE. Unfortunately for Ruska, his preferred successor, Wolfgang Dieter Riecke, left Dahlem in 1970 headed for Baden-Baden, where he had been promised his own institute and a high-performance microscope set up on the rocky, vibration-free floor of the Black Forest. Riecke was of the opinion that atomic-resolution electron microscopy would be, in principle, impossible in Berlin because of the intense ground vibrations.

Ruska was disappointed but not discouraged and still had enough clout to fight for the IFE. He attempted to groom as his new successor Karl-Heinz Herrmann, whom he had picked up from the Siemens research department in 1971. Furthermore he secured financing for construction of a new electron microscopy building that would be completed in 1974. The building included two double-walled towers in which two DEEKO electron microscopes, mounted in steel cages, could be suspended from above by cables and thereby kept essentially free from vibration. The microscopes were securely fastened to the weighty metal cages, while the observer stood on a floor projecting from the tower walls and was thus effectively removed

Fig. 5.12. *Helmut Baumgärtel (left) and Heinz Gerischer, 1994.*

the insular character of West Berlin – and island in the red sea, as the folkore of the time had it. But the FHI and BESSY would come to share something much more intimate than a "preferred user" relationship. In light of the special interest in synchrotron radiation shown by Bradshaw and Gerischer, the CPT Section agreed to a proposal from the Board of the FHI in February of 1978 that the Scientific Director of BESSY should also be appointed Scientific Member of the FHI. A search committee formed within the CPT Section to consider exclusively this appointment, separate from the committee already in place to discuss the future of the Institute. After considering several possible candidates, the Board of the Institute recommended Harald Ibach of the then Jülich Nuclear Research Center, but Ibach declined the offer. Further attempts to attract a suitable external candidate similarly faltered, and the Board soon turned its attention to Bradshaw. However, Helmut Baumgärtel would become the first Scientific Director of BESSY and Gottfried Mülhaupt, who was responsible for building the storage ring itself, its first Technical Director. Bradshaw then took over the post of Scientific Director at the beginning of 1981, eighteen months before the facility began normal operations, and just after he was promoted to Scientific Member of the FHI and Director of the Department of Surface Physics. Bradshaw remained Scientific Director at BESSY until the end of 1985, then returned to the post for roughly a year in 1988, following the death of his successor Ernst-Eckhard Koch. In addition, Anselm

Stieber, Executive Administrator of the FHI, also served as Administrative Director of BESSY, but this arrangement proved unsatisfactory and was abandoned after 1984. The success of user operation at BESSY in the early years was due mainly to a small, but extremely capable in-house group, headed by William Peatman.

The challenges of administering BESSY occupied much of Bradshaw's time during the early 1980s. He also faced some struggles with regard to funding and prestige in light of the unusual nature of his promotion, i.e. his remaining at least partly within the institute from which he was promoted. Nevertheless, Bradshaw and his FHI group managed to make several significant scientific contributions during the 1980s, in particular to synchrotron instrumentation. Together with Eberhard Dietz and Walther Braun, Bradshaw built a high-flux, high-energy toroidal grating monochromator (HE-TGM-1) for BESSY that began operation in 1984. In collaboration with Manuel Cardona, among others, Bradshaw also developed a VUV ellipsometer that enabled novel researches into the optical properties of solids and surfaces. The first experiments with the infrared component of synchrotron radiation, to which Erhard Schweizer and Ernst Lippert were key contributors, were also made at this time. At the close of the decade, the Bradshaw group, in particular Josef Feldhaus, then took a leading role in the construction of the X 1B undulator beamline at the National Synchrotron Light Source in Brookhaven, New York. This project foreshadowed Bradshaw's efforts in the 1990s, both in furthering the utilization of undulator radiation and in the experiments it made possible, particularly in molecular photoionisation.

Fig. 5.13. *BESSY I in Berlin-Wilmersdorf, 1986.*

Even before Bradshaw's first term as scientific administrator at BESSY came to an end, he had begun lobbying for the construction of a new "third-generation" synchrotron radiation source in Berlin. Since the construction of BESSY (hereafter BESSY I), physicists had developed "wigglers" and "undulators" that could increase the spectral brilliance of synchrotron radiation several orders of magnitude by inducing periodic, "sideways" oscillations of the electron beam in otherwise straight sections of the storage ring. Just three years after BESSY I became operational, Bradshaw and colleagues Gottfried Mülhaupt, William Peatman, Walter Braun and Franz Schäfers sent a proposal to the BESSY Supervisory Board for a 1.5 GeV storage ring at the site in Berlin-Wilmersdorf using BESSY I as injector.[48] The idea behind "BESSY II" was to cover roughly the same soft X-ray spectral range as the BESSY I bending magnets but with the much brighter undulator radiation. By February of 1986 interest in the proposed storage ring was widespread enough to attract 26 leading scientists and engineers to a preliminary planning meeting at BESSY I. Two further planning meetings followed later in the same year, the second of them held at the Fritz Haber Institute. The first published proposal for BESSY II appeared at the end of the year and included among its contributors not only Alexander Bradshaw and Ernst-Eckhard Koch, but also Karsten Horn, Dieter Kolb and Josef Feldhaus of the Fritz Haber Institute and Hans-Joachim Freund, who was then at Erlangen but would join the FHI as Director of the Department of Chemical Physics in 1996. The proposal encountered some difficulties, as the potentially available land near BESSY I was devoted to horticultural activities; even an underground site met with opposition. But, in 1989, when the Berlin Wall fell and reunification quickly followed a whole host of new sites became available. A site was chosen in Adlershof, and the plan to use BESSY I as an injector was abandoned.[49] Instead, after BESSY II had gone into operation in 1998, BESSY I was disassembled and sent to Jordan as the starting point for the SESAME project. No parallel appointments similar to those spanning the FHI and BESSY I were made, but the ties between synchrotron radiation sources and research at the FHI remained firmly intact, above all through the efforts of Bradshaw and members of his Department of Surface Physics, but also through the work of Hans-Joachim Freund and Robert Schlögl and their respective Departments (cf. Chapter 6).

In addition to experiments at the FHI in surface vibrational spectroscopy, with Brian Hayden and Horst Conrad, and in low temperature STM with Erhard Schweizer, Beat Briner and Hans-Peter Rust, Bradshaw embarked upon three lines of research that specifically took advantage of the availability of synchrotron radiation sources. The first, a long-running application of "energy scan" photoelectron diffraction to the study of adsorbed molecules and molecular fragments in collaboration with Phillip Woodruff of the University of Warwick, made extensive use of the BESSY facilities and earned Bradshaw and Woodruff the Max Planck Research Prize in 1994. Although photoelectron diffraction had

---

48 Bradshaw, Gaupp, Koch, Maier, Peatman, *BESSY II*.
49 Wista-Management, *Adlershof*.

Fig. 5.14.   *BESSY II in Berlin-Adlershof.*

been known for more than 15 years when Bradshaw and Woodruff began their collaboration, they were able to provide novel quantitative structural information for over a hundred adsorption systems (to date) by taking full advantage of synchrotron radiation and of efficient, innovative simulation codes written by Volker Fritzsche. So-called direct methods were also pioneered in the group at this time, particularly by Philip Hofmann. The second line of research undertaken by Bradshaw and his colleagues involved a series of photoionization studies of free molecules in the core-level region at hitherto unavailable spectral resolution. This research made extensive use of the previously mentioned X1B beamline at Brookhaven. Among other novelties, these investigations demonstrated the importance of vibronic coupling in molecules containing equivalent cores and provided the first measurements of the influence of shape resonances and double excitations on the vibrational fine structure of core level lines of various small molecules. Bradshaw's third major area of research in the 1990s was low energy electron microscopy and photoelectron microscopy pursued in collaboration with Winfried Engel and Elmar Zeitler of the Department of Electron Microscopy. After extensive laboratory studies on, amongst other things, reaction-diffusion fronts in heterogeneous reactions, this work resulted in a proposal for a photoelectron spectro-microscope for BESSY II (the SMART project) presented together with Hans-Joachim Freund and with Eberhard Umbach, then of the University of Würzburg and later External Scientific Member of FHI.

These researches were, however, only the latest expressions of a long-standing tradition of pursuing atomic and molecular physics using synchrotron radiation within the Department of Surface Physics. The group of Ulrich Heinzmann performed pioneering studies of photoionization processes using circularly polarized radiation and spin-polarized detection of the photoelectrons, before Heinzmann was appointed to a Chair in Bielefeld in 1985. Ernst-Eckhard Koch, Heinzmann's

Fig. 5.15. *Experimental Hall of BESSY I, 1995.*

successor at the FHI, also undertook seminal experiments on molecular crystals before his untimely death in 1988. Koch was succeeded by Uwe Becker, then at the TU Berlin, who has since performed numerous pioneering experiments on the dynamics of photoionization processes and now leads a group in the Department of Molecular Physics. A separate photoionization group, headed by Uwe Hergenhahn, would accompany Bradshaw to Garching (see below).

These research activities had to compete for time with growing demands for Bradshaw's skills as a scientific administrator. Bradshaw served only one term as Executive Director of the FHI, from 1990 to 1991. At the same time, he was elected President of the Berlin chapter of the German Physical Society, and in 1996 he was appointed to the Executive Committee of the German Physical Society (DPG) and later elected President of the Society, beginning a two-year term of office in 1998. While DPG President, he was the co-initiator of the Year of Physics in 2000, the first "Science Year" in Germany. 1998 also saw the launch of *New Journal of Physics*, an open access, peer-reviewed journal which Bradshaw co-founded with colleagues from the British Institute of Physics and for which he later served as the first Editor in Chief. But that same year, the Max Planck Society called upon Bradshaw to take over as Scientific Director of its Institute for Plasma Physics (IPP) in Garching. On the advice of Hubert Markl, then President of the MPG, Bradshaw did not immediately relinquish his directorship at the FHI, since the political climate in Germany appeared somewhat hostile to the funding of nuclear fusion research, in which the IPP specializes. But by the end of 2001 Bradshaw had fully invested himself in his new post and decided to "stick with [the IPP] through thick and thin." Bradshaw stepped down as Department Director at the FHI and the remaining Directors began in earnest their search for his successor. The Board had in mind from the outset an appointment that would go beyond and yet complement the strengths of the FHI in surface science and catalysis research. The result was the appointment of Gerard Meijer in 2002 as head of a Department of Molecular Physics. Bradshaw, however, maintained ties with the FHI and BESSY and, after relinquishing his post at the IPP at the end of 2008, he returned to the FHI and rededicated himself to research on the photoionization of molecules and clusters.

Fritz Haber Institute as an International Center for Surface Science

Simultaneous with Bradshaw's efforts, other significant changes were taking place at the Institute that would transform it into an international hub of surface science. Deliberations about creating the Institute's own theory department were already a part of the 1980 restructuring effort. The idea was to strengthen the FHI in the area of theory at the level of an Institute Director. A lengthy search identified Matthias Scheffler as a candidate for the directorship, and the MPG Senate called Scheffler to became Scientific Member and Director of the Theory Department on 1 July 1988. Another crucial facility for the FHI has been the Joint Network

Fig. 5.16.   *The "old" FHI computer center with the memory units of the DECSYSTEM-2020, 1974.*

Fig. 5.17. *Gerhard Ertl (left) and Heinz Gerischer, 1981.*

Center, created in 1987. Its forerunner, the "old" computer center, led by Jürgen Kühn, was incorporated into this new center, and the workhorse of the old center, the Digital Equipment Corporation's DECSYSTEM-2020 computer, was passed on to the MPI for Cell Biology in Ladenburg. The Joint Network Center has served all of the Berlin MPIs ever since.

Parallel to these efforts on the theory side were efforts to further strengthen experiment at the Institute. In 1983, Elmar Zeitler, then Executive Director of the FHI, traveled to Munich to sound out whether Gerhard Ertl might accept an offer from the MPG to become a director at the FHI; this was an unlikely outcome in the eyes of many. The University of Munich offered Ertl generous support for his research, and he was and remains a strong supporter of the universities and their students. However, Ertl accepted the offer, in part guided by what he later called an "emotional reason:" coming to the FHI enabled him to become the successor of his mentor Heinz Gerischer, who remained a scientific role model for Ertl. Moreover, his experiences in Stuttgart helped convince Ertl that an "optimal arrangement" was possible, in which a Max Planck Director works extensively with doctoral students and junior researchers, contributing in this way to the mission of the universities, but remains free of routine teaching and administrative duties. It was this arrangement that Ertl set out to establish in Berlin.

Ertl was appointed Scientific Member of the MPG and Director at the FHI as of 1 April 1985. Rather than taking over from Gerischer, he shared in directing the Physical Chemistry Department for the next two years with Gerischer, commuting between Berlin and Munich, to which he was still bound by previous commitments. The smooth transition from Gerischer to Ertl was completed with Gerischer's retirement on 1 April 1987, at which point Ertl became the sole director of the Department.

The year 1986 at the Institute was shaped by two events. On 15 October 1986, the news arrived from Stockholm that Ernst Ruska had been awarded a Nobel Prize in Physics "for his fundamental work in electron optics, and for the design of the first electron microscope." Ruska shared the prize with the inventors of the scanning

Fig. 5.18.   *Ernst Ruska (left) and Elmar Zeitler at the departmental Nobel Prize celebration on 30 October 1986 in Ernst-Ruska-Bau.*

tunneling microscope, Gerd Binning and Heinrich Rohrer. The festivities at the Institute took place on 30 October 1986. First, there was an informal gathering of Ruska's former coworkers and members of the Department of Electron Microscopy in the 1974 building, which had been renamed after Ruska shortly before the arrival of the Stockholm news. This was followed by a gala reception, given by the Land of Berlin, in the large lecture hall of the Institute, which was attended by numerous prominent figures from science and politics, including the Mayor of Berlin Eberhard Diepgen, the MPG President Heinz A. Staab and the Physics Nobel Laureate from the preceding year Klaus von Klitzing.

The festivities did not fall accidentally on 30 October 1986. On the next day, the 75th anniversary of the founding of the Institute was celebrated in the main lecture hall of Freie Universität; in the evening, there was a follow-up reception at the Harnack House. At the ceremony, Heinz Staab extolled the significance of the Fritz Haber Institute for the Max Planck Society and Heinz Gerischer, in a keynote address, surveyed the history of the FHI and its ongoing research endeavors. Historian Fritz Stern of Columbia University, whose family had been friends of Fritz Haber and his family in Breslau, spoke on the topic "Fritz Haber in War and Peace." In his talk, Stern also grappled with a brochure, published in connection with the anniversary of the Institute by a group consisting of members of the Institute and young historians,[50] which leveled criticisms against the Institute and against Haber as the father of chemical warfare. In this way, the political turbulences of the

---

50   Chmiel, Hansmann, Krauß, Lehmann, Mehrtens, Ranke, Smandek, Sorg, Swoboda, Wurzenreiner, *Bemerkungen.*

Fig. 5.27. *An outing of Ertl's Department in Berlin-Tegel, 1990s.*

Ertl's research has demonstrated that a rather simple system (a chemical reaction occurring between two diatomic molecules on a well-defined single crystal surface with fixed external parameters and well-established mechanism) can be used to study (and model) a quite complex behavior. The conclusions which it allows us to draw about far-from-equilibrium open systems transcend catalysis and surface science and provide clues about laws believed to govern the whole of nature.

However, during the 1990s, there were more than just scientific challenges with which to grapple. The fall of the Berlin Wall and the ensuing German reunification had, of course, greatly impacted the Fritz Haber Institute and indeed the entire Max Planck Society. Due to the special political status of West Berlin, there were few personal contacts and almost no institutionalized scientific contacts with colleagues in the GDR. An exception was the Leopoldina Academy in Halle, whose members, from the East and the West, could see each other at Leopoldina's annual meetings. Ernst Ruska and Heinz Gerischer, both of whom had personal ties to East Germany, vigorously cultivated the Leopoldina connection. Ernst Ruska in particular nurtured a collegial relationship with the director of the Central Institute for Electron Microscopy and Solid-State Physics of the Academy of Sciences of the GDR in Halle, Heinz Bethge, who became Leopoldina's President in 1974. Their relationship was strengthened, for instance, through the publication in 1979 of Ruska's account of the history of electron microscopy[55] in the series Nova Acta Leopoldina, on the one hand, and through Bethge's membership on the FHI's Advisory Board (a highly unusual arrangement at the time) on the other. As for the direct counterpart of the FHI in the East, the Central Institute for Physical

---

55 Ruska, *Elektronenmikroskopie.*

MPG and the German Reunification

MPI for Microstructure Physics, Halle

The Unification Treaty between the Federal Republic of Germany and the German Democratic Republic from the Summer of 1990 and the negotiations that preceded it foresaw a unified research landscape for a reunified Germany. Its structure was to emulate that of the old Federal Republic. For the Max Planck Society this implied that it would assume the same role in the new Federal Republic as it held in the old one.

In late Fall of 1989, the MPG put in place an informal cooperation program to foster mutual visits and joint projects between researchers in the East and West of Germany. In the Summer of 1990, this was replaced by a program aimed at establishing special MPG Research Groups. These would be set up on the initiative of a Max Planck Institute, which would then also manage them. The MPG Research Groups were intended to enable an outstanding GDR researcher to take advantage of the superb research facilities and resources available at a Max Planck Institute and to embed him or her in the international scientific community. The research groups were funded by the MPG but based at the universities in the East, since the primary goal was to strengthen research at these universities. During 1991-92, the MPG established, in three installments, 29 Research Groups, whose subjects ranged from gravitational physics (University of Jena) to complex catalysis (University of Rostock), enzymology of peptides (University of Halle), structural grammar (Humboldt University) and estate ownership East of the Elbe (University of Potsdam); in addition, seven research centers in the humanities were founded. The funding was typically limited to a period of five years, and thereafter the MPG Research Groups were integrated into the universities.

In a parallel initiative, the Sections of the MPG set up panels tasked with developing ideas for establishing new project groups and institutes in the new Lands. Particular attention was given to any innovative, high-performance research institutions of the former GDR. In the end, only the Institute for Electron Microscopy and Solid-State Physics of the Academy in Halle, as well as parts of the Institute for Polymer Chemistry in Teltow-Seehof, were recommended for integration as Max Planck Institutes. In the process of appointing their new directors, brand new research institutes, the MPI for Microstructure Physics and the MPI of Colloids and Interfaces, were established. The establishment of these and 18 other Max Planck Institutes in the new Lands during the 1990s followed the general rules and procedures of the MPG; also here the Harnack principle was applied in order to enable outstanding scholars to pursue research on important topics that showed great promise.

With a total of 20 Max Planck Institutes established in the new Lands, the density of the Max Planck Institutes has become roughly even throughout the reunified Germany.

Chemistry of the Academy of Sciences of the GDR in Berlin-Adlershof, there was essentially no official contact. In the early 1980s, Block had attempted, in vain, to initiate such contact. In the late 1980s, members of the Physical Chemistry Department, in particular Hellmut Karge, cultivated contacts with colleagues at the University of Leipzig, above all Harry Pfeifer and his NMR-spectroscopy group; there were also contacts to the department of statistical thermodynamics and theoretical biophysics at Humboldt University, as their work was relevant to Ertl's research on non-linear dynamics of surface processes. There was correspondence between Ertl and Werner Ebeling, the head of the department, as well as visits by Ertl's coworkers at conferences and workshops organized by Ebeling's group. Since West Berlin represented, in the eyes of the GDR, a "peculiar political entity" official contact with West Berlin institutions was significantly more difficult to set up than with institutes in other Western countries or even West Germany. Similarly, from the western side, West Berlin's political status implied certain restrictions. As a result, not even "bilateral" relations between institutions or individuals could be officially established. Instead, contacts were cultivated in part semi-officially and in part privately, in the case of the Ebeling group mainly by younger members of the FHI. Alexander Bradshaw, as a British citizen, was in a somewhat less tenuous position and so became the most active among his FHI colleagues in establishing relations with GDR academia. He was able to visit colleagues at the University of Leipzig and launch a modest cooperation between Eastern colleagues and BESSY

Fig. 5.28. *Heinz Bethge (left) bestowing the Cothenius Medal of the Leopoldina on Ernst Ruska, Halle 1975.*

in the area of X-ray spectroscopy. He also managed to take part in conferences of the Surface Physics division of the Physical Society of the GDR and to give lectures in Jena, Leipzig and Dresden, as well as at the Adlershof Academy Institutes of Optics and Physical Chemistry. The political delicacy of such events is illustrated by the fact that Bradshaw's Adlershof audience was hand-picked, and no member of the institute management was in attendance, nor did they receive Bradshaw at any time during his visit.[56] On political grounds, such visits were handled on the lowest level of protocol, confined to the grey area of the semi-official. That is until the fall of 1987, when a scientific cooperation treaty established a legal framework for German-German scientific relations.

Two years later, such ploys became irrelevant. The fall of the Wall and the democratization of the GDR erradicated all restrictions, and the previously detrimental island position of West Berlin, became a boon for rapidly establishing personal contacts between scientists from the West and the East. However, the problems of the day remained formidable, as the German reunification required not only a reform and restructuring of the academic research establishment of the GDR but indeed its full integration into the Federal Republic.

The FHI was an active participant in this integration process. At the outset, the FHI invited certain GDR scientists to come to the Institute to either familiarize themselves with FHI research or to carry on with their own research using FHI facilities. Alexander Bradshaw and Gerhard Ertl were members of the evaluation

Fig. 5.29.   *The Senate of the MPG in front of the main entrance of the FHI, 22 May 1981. Occasioned by the Annual Meeting of the MPG that year in Berlin.*

---

56 *Sehr visionär und kühn*, p. 24.

Theory Department

Since 1987, computational physics and chemistry have been pursued at the Institute by Matthias Scheffler and his collaborators. Their principal aim has been to develop a predictive theory of materials, one capable of aiding in the design of new and the improvement of existing materials, in particular those materials used in semiconductor physics and in heterogeneous catalysis on metal and oxide surfaces. To be relevant to material design the theory needs to allow researchers to model materials subject to real-life environments, e.g. appropriate gas mixtures under realistic pressures and temperatures. Such a predictive theory requires a combination of an accurate description of atomic-scale processes with a treatment of their statistical interplay at meso- and macroscopic scales, which controls phenomena such as steady-state crystal growth and heterogeneous catalysis. This, in turn, requires attention to the behavior of matter on diverse scales, both spatial (from fractions of a nanometer to a millimeter) and temporal (from femtoseconds to hours), see Fig. 6.2. Among other results and insights, the work of the Theory Department has demonstrated that a "one-structure, one-site, one-mechanism" description is, in general, inadequate for understanding the function of materials at authentic ambient conditions.

The treatment of the atomic-scale processes is based on *ab initio* electronic structure theories suitable for treating both valence and dispersion forces. This entails tackling the many-body, i.e. many-electron problem, which requires the use of approximation techniques. The method of choice for Scheffler and coworkers is the density functional theory (DFT), whose development was initiated by Walter Kohn and coworkers circa 1964–65. Greatly enhanced by recent theoretical and

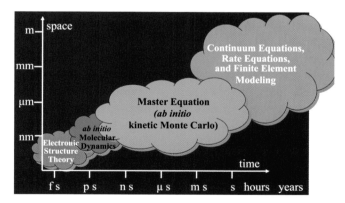

Fig. 6.2. *Temporal and spatial scales involved in materials science applications, such as in heterogeneous catalysis. The elementary processes of bond breaking and bond making between atoms and molecules are described by the electronic-structure theory, from which the rest unfolds. The function of a catalyst is determined by an interplay among many molecular processes and only develops over meso- and macroscopic lengths and times.*

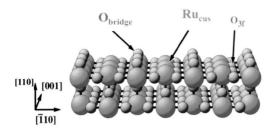

Fig. 6.3.   *Surface structure of RuO₂ (110) under UHV conditions as predicted by DFT calculation and observed experimentally. If we ignore relaxations this is essentially a truncated bulk geometry. All O$_{bridge}$ sites are occupied, and all Ru$_{cus}$ sites are empty (cus = coordinatively unsaturareted site). Also shown are the second layer three-fold (3f) coordinated O atoms.*

computational advances, density-functional theory can now provide 'chemical accuracy' for sizable systems that contain hundreds of gaseous or condensed-phase atoms. Developing computational methods and computer codes that enable this new level of accuracy is among the core activities of the Theory Department.

On the next level of the hierarchy laid out in Fig. 6.2, electronic structure theory is linked to molecular-dynamics simulations that include energy dissipation and non-adiabaticity. A reduction of the computational cost of the electronic structure calculation per time step makes simulations over a longer time scale feasible without compromising accuracy. Simulations that extend over tens of picoseconds could thus be implemented for rapid processes in gases and solids, at interfaces and in solutions. An alternative approach, commonly used in molecular dynamics simulations, is based on empirical force fields. This approach often overlooks features that are critical to chemical reactions occurring in realistic probes and devices; although, "*ab initio*-on-the-fly" interatomic forces offer one possible remedy. Even though massively parallel computers alleviate some of the length scale problems and many of the accuracy issues encountered in this work, they are not necessarily beneficial in tackling the associated time scale problems, since time integration is sequential. Often a more effective way of dealing with time scale challenges is to employ long-term dynamics methods founded on the master equation of statistical mechanics. Numerical implementation of these methods, when linked with density-functional theory, is referred to as *ab initio* kinetic Monte Carlo (kMC), and is based on a set of rate constants calculated atomistically at the molecular dynamics level and tested for completeness via extensive molecular dynamics runs. As a result, the *ab initio* kinetic Monte Carlo approach has reached the point where processes can be tackled over time scales ranging from seconds to hours.

However, beyond the scale of molecular dynamics and kinetic Monte Carlo the links to the mechanics of continua and to the rate equations become tenuous. Establishing robust links to these higher levels is a major challenge.

One concrete example of the research methods just described, is the multi-scale modeling of CO oxidation on the Ru/RuO₂ catalyst at realistic ambient conditions and the resulting description of the steady state of the catalytic conversion process. CO oxidation is a strongly exothermal reaction. Yet, in the gas phase, the CO + $\frac{1}{2}O_2 \leftrightharpoons CO_2$ reaction is spin-forbidden, since the reactants have a total spin $S = 1$ (due to the triplet ground state of O_2) while the reaction product (CO_2) has zero spin, $S = 0$. However, for dissociated O_2, the individually adsorbed O atoms are in a spin zero state and so the reaction becomes spin-allowed. The reaction rate depends on an energy barrier determined by the properties of the adsorbent. But, obviously the adsorbed CO and O species must also occupy nearby positions. Thus, in order to predict the turn-over frequency, i.e. the number of CO_2 molecules formed per unit area of the catalyst's surface per unit time, one needs to take an appropriate statistical average over space and time.

Predictive modeling of heterogeneous catalysis must be able to represent the steady state of operation. Here the statistical mechanics of the various mutually interfering atomistic processes reveals the significance of instabilities and fluc-tuations; a catalyst is a "live" system, subject to incessant changes even in its steady state. As it turns out, these instabilities and fluctuations are crucial for the self-healing of locally poisoned areas of the catalyst and, hence, for its long-term operational stability and are particularly prominent at conditions of steady-state, high catalytic performance.

Catalysis on the Ru metal occurs at ambient conditions (pressure and temper-ature) where the bulk oxide, RuO_2, is stable. However, this does not reveal much about the composition and structure of the surface. Under ultra-high vacuum conditions, the surface composition and structure of RuO_2 has been predicted by the density-functional theory, Fig. 6.3. Here the surface is nearly perfect, with

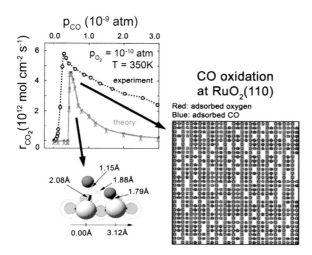

Fig. 6.4. *Calculated and measured CO_2 turn-over rates as a function of CO pressure (top left); a snapshot of the surface occupancy under the conditions of optimal catalytic perfor-mance in the course of a kMC simulation (right); geometry of a transition state of the CO oxidation reaction (bottom left).*

almost no vacancies or other defects in the O-bridge rows and no adatoms at the coordinatively unsaturated sites (cus), Ru_{cus}. This prediction is in excellent agreement with scanning-tunneling microscopy results.

Moving pictures of the detailed atomic structure together with a "sensitivity analysis" reveal the importance of kinetics for understanding the conditions for high-performance catalysis: The dissociative adsorption of O_2 and the non-dissociative adsorption of CO compete for adsorption sites on the surface, specifically for the bridge and the coordinatively unsaturated sites, see Fig. 6.4. Here an important correlation is noted, owing to the fact that O_2 requires two nearby sites while CO needs just one. Hence after the catalytic reaction, $O_{ad} + CO_{ad} \rightarrow CO_2$, has taken place, two empty sites are left behind. These can be occupied by two O adatoms (from the dissociated O_2) or by CO. However, as soon as one CO molecule has been adsorbed, the remaining empty site is no longer conducive to O_2 dissociation and so can only accommodate another CO molecule. As a consequence, kinetically-controlled, non-random structures form that disable some potentially active reaction pathways. In fact, the whole surface may become excessively CO-rich and thus catalytically inactive. Only if the pressure is chosen properly can the adsorbed CO desorb again at a rate sufficient to "heal" such "poisoned" regions. Alternatively, the surface may become O-rich, in which case it can only be healed if some desorption of oxygen takes place and/or chemical reactions erode away these regions from the edges. The calculated turn-over frequency of CO to CO_2 conversion as a function of O_2 and CO pressures identifies conditions where the steady state changes from an "O-poisoned" to a "CO poisoned" surface, see Fig. 6.5. The optimum, high performance conditions are found in between. In the active regime, the surface proceeds locally from an O-rich or a CO-rich to a catalytically highly-active composition and back to an inactive one. This is a clear-cut example of the sort of fluctuation and structural instability, referred to earlier in general terms, that is vital for sustained catalytic performance.

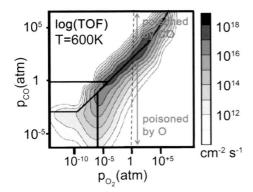

Fig. 6.5.   *Map of calculated turn-over frequencies (TOFs) at a temperature of 600 K for the oxidation of CO on the Ru/Ru₂ catalyst. The plot is based on 400 kinetic Monte Carlo simulations for different CO and O_2 pressures.*

The Director of the Theory Department is Matthias Scheffler, born in 1951 in Berlin. He earned his PhD in Physics from the Technische Universität Berlin in 1978 with a thesis written at the FHI on "Theory of Angular Resolved Photoemisson from Adsorbates." His advisers were Kurt Molière (FHI), Kyozaburo Kambe (FHI) and Frank Forstmann (Freie Universität Berlin and FHI). Prior to his appointment as Director at the FHI in 1988, he was a staff scientist at the Physikalisch-Technische Bundesanstalt in Braunschweig. In 2004 he was appointed "Distinguished Visiting Professor for Computational Materials Science and Engineering" at the University of California Santa Barbara, where he spends up to a quarter of the year.

The following researchers previously or currently affiliated with the Theory Department over periods of several years had a noticeable impact on the Department's work:

- Volker Blum (PhD in Physics 2001, Universität Erlangen-Nürnberg, adviser Klaus Heinz; at FHI since 2004)

- Jarek Dabrowski (PhD in Physics 1989, Polish Academy of Sciences, adviser Tadeusz Figielski; at FHI 1989–1992; at present staff scientist at Leibniz-Institut für innovative Mikroelektronik, Frankfurt/Oder, Germany)

- Kristen Fichthorn (PhD in Chemical Engineering 1989, University of Michigan, advisers Robert Ziff and Erdogan Gujlari; at FHI 1998–1999; at present Professor at The Pennsylvania State University, University Park, USA)

- Martin Fuchs (MSc in Physics 1992, Oregon State University, adviser Philip Siemens; at FHI 1992–1999, 2002–2009; at present an official at the European Parliament in Brussels)

- Veronica Ganduglia-Pirovano (PhD in Physics 1989, Universität Stuttgart, adviser Peter Fulde; at FHI 1994–2001; at present Professor at the Universidad Autónoma de Madrid)

- Xavier Gonze (PhD in Applied Sciences/Engineering 1990, Université Catholique de Louvain, adviser Jean-Pierre Michenaud; at FHI 1998–1999; at present Professor at Université Catholique de Louvain, Louvain-la-Neuve, Belgium)

- Axel Groß (PhD in Physics 1993, Technische Universität München, adviser Wilhelm Brenig; at FHI 1993–1998; at present Professor at University of Ulm, Ulm, Germany)

- Bjørk Hammer (PhD in Physics 1993, Technical University of Denmark; adviser Karsten Jacobsen/Jens Nørskov; at FHI 1993–1994; at present Professor at Aarhus University, Aarhus, Denmark)

- Klaus Hermann (PhD in Physics 1974, Technische Universität Clausthal, advisers Ernst Bauer and Lothar Fritsche; at FHI since 1990)

- Joel Ireta Moreno (PhD in Physical Chemistry 1999, Universidad Autónoma Metropolitana-Iztapalapa, adviser Marcelo Galván; at FHI 1999–2007; at present Professor at Universidad Autónoma Metropolitana-Iztapalapa, Mexico City, Mexico)

- Hong Jiang (PhD in Physical Chemistry 2003, Peking University, adviser Xinsheng Zhao; at FHI 2006–2009; at present Professor at Peking University, Beijing, China)

- Kyozaburo Kambe (PhD in Physics 1962, Freie Universität Berlin, adviser Kurt Molière; at FHI 1956–1991)

- Peter Kratzer (PhD in Physics 1993, Technische Universität München, adviser Wilhelm Brenig; at FHI 1997–2006; at present Professor at Universität Duisburg-Essen, Duisburg, Germany)

- Sergey Levchenko (PhD in Chemistry 2005, University of Southern California, adviser Anna I. Krylov; at FHI since 2008)

- František Máca (PhD in Physics 1983, Czechoslovak Academy of Sciences, adviser Igor Bartoš; at FHI 1988–1991; at present postdoctoral researcher at Academy of Sciences of the Czech Republic, Praha, Czech Republic)

- Michael Methfessel (PhD in Physics 1986, Katholieke Universiteit Nijmegen, advisers Jürgen Kübler and Alois Janner; at FHI 1989–1992; at present staff scientist at Leibniz-Institut für innovative Mikroelektronik, Frankfurt/Oder, Germany)

- Angelos Michaelides (PhD in Chemistry 2000, Queen's University Belfast, adviser Peijun Hu; at FHI 2003–2006; at present Professor at University College London, London, United Kingdom)

- Jörg Neugebauer (PhD in Physics 1989, Humboldt Universität Berlin, adviser Rolf Enderlein; at FHI 1990–1993, 1996–2003; at present Director at Max-Planck-Institut für Eisenforschung GmbH, Düsseldorf, Germany)

- Oleg A. Pankratov (PhD in Physics 1977, Moscow Physical-Technical Institute, adviser Evgeniy Maksimov; at FHI 1990–1995; at present Professor at Universität Erlangen-Nürnberg, Erlangen, Germany)

- Eckhard Pehlke (PhD in Physics 1989, Christian-Albrechts-Universität Kiel, adviser Wolfgang Schattke; at FHI 1991–1996; at present Professor at Christian-Albrechts-Universität Kiel, Kiel, Germany)

- Christian Ratsch (PhD in Physics 1994, Georgia Institute of Technology, adviser Andy Zangwill; at FHI 1995–1997, 2003; at present Professor at University of California, Los Angeles, USA)

- Karsten Reuter (PhD in Physics 1998, Universität Erlangen-Nürnberg, adviser Klaus Heinz; at FHI 2003–2009; at present Professor at Technische Universität München, München, Germany)

- Patrick Rinke (PhD in Physics 2002, University of York, adviser Rex Godby; at FHI since 2003)

- Angel Rubio (PhD in Physics 1991, University of Valladolid, adviser Carlos Balbas; at FHI (distinguished visiting scientist) since 2009; at present Professor at the University of the Basque Country, Donostia-San Sebastian, Spain)

- Paolo Ruggerone (PhD in Physics 1989, University of Milan, adviser Giorgio Benedek; at FHI 1994–1998; at present Professor at Università deglo Studi di Cagliari, Cagliari, Italy)

- Arno Schindlmayr (PhD in Physics 1999, University of Cambridge, adviser Rex Godby; at FHI 1998–2003; at present Professor at Universität Paderborn, Paderborn, Germany)

- Catherine Stampfl (PhD in Physics 1990, La Trobe University, Melbourne, adviser John D. Riley; at FHI 1990–2003; at present Professor at University of Sydney, Sydney, Australia)

- Alexandre Tkatchenko (PhD in Physical Chemistry 2007, Universidad Autónoma Metropolitana, Iztapalapa, adviser Marcelo Galván; at FHI since 2007)

- Chris van de Walle (PhD in Electrical Engineering 1986, Stanford University, adviser Richard M. Martin; at FHI 1999; at present Professor at University of California, Santa Barbara, USA)

- Byung Deok Yu (PhD in Physics 1992, Seoul National University, adviser Jisoon Ihm; at FHI 1994–1997; at present Professor at University of Seoul, Seoul, Korea)

Department of Inorganic Chemistry

Since its inception in 1994, the Department of Inorganic Chemistry has advanced a research program focused on studying, and bridging, the gap between catalytic model systems of surface science and real heterogeneous catalysts. Originally housed on the premises of Elmar Zeitler's Department of Electron Microscopy, the Department relocated five times before settling in the refurbished former IFE building, which provides it adequate space to house facilities ranging from high-resolution electron microscopes to synthetic-chemistry laboratories.

The current, functional understanding of heterogeneous catalysis relies on the work of Gerhard Ertl and Gabor Somorjai, who pioneered the single-crystal approach to modeling a metal catalyst. This approach led to a quantitative description of elementary processes relevant for catalytic transformations which, in turn, became a pre-requisite for the development of a theory of heterogeneous catalysis. However, despite the solid conceptual framework established by the theory, it remains nearly impossible to predict the catalytic properties of working catalysts developed by empirical or combinatorial methods. Finding the missing link between the conceptual and practical understandings of heterogenous catalysis has become the long-term goal of the Department.

At the core of present research is analytic work aimed at determining the essential ingredients of the high-performance, complex catalytic systems encountered in chemical technology. This analysis is followed by a synthesis of the identified functional materials whose kinetic characteristics match those of the complex catalytic system. In order to probe the catalytic activity of the synthesized materials, the Department has developed a suite of in situ analytical techniques that mimic real-life conditions. This enables Department members to establish quantitative relationships between the structure and the function of the materials involved.

Establishing such relationships is complicated by the dynamic response of a catalyst to its reaction environment, which leads to the formation of metastable surfaces containing the active sites as minority features. The active catalytic sites change their structure in a cycle which brings them back to their initial state after the completion of each reaction sequence. Since the sites are metastable with respect to the equilibrium state of the bulk catalyst phase, the catalyzed chemical reaction can proceed beyond its equilibrium. But if the cycles are interrupted by separating the catalyst from its reaction environment, for the purpose of analysis or static experimentation, the metastable active sites decay to stable products, thereby precluding their structural identification.

This creates gaps of understanding between surface science and the performance of complex catalytic systems. Indeed, the origin of the failure to predict the catalytic properties of complex systems lies in the conundrum that catalysts can either be rigorously analyzed in non-functioning forms or operated at high performance without knowing their active structure. In order to bridge this gap, Department members synthesize functioning models with minimal complexity and investigate their structural properties in situ. The Institute's rich background in catalysis and its focus on model systems (*cf.* the Department of Chemical Physics) facilitate such interdisciplinary research on the interface between physical and inorganic chemistry. The development of the Department of Inorganic Chemistry can be characterized as a gradual adaptation to such an interdisciplinary effort – using the opportunities of the Institute's existing infrastructure on the one hand and providing stimuli for the work of other Departments on the other.

When the Department was founded, it had already been realized that electron microscopy of catalysts could provide unique access to the non-translational aspects of their structure, but taking advantage of this realization required a full arsenal of microscopic techniques and cutting-edge instrumentation. Between 1995 and 2007, the Department replaced its outdated microscopes with a set of modern instruments. The know-how of Zeitler's Department in constructing and operating electron microscopes has thus contributed to novel insights into the structure of real polycrystalline catalysts. The method of electron energy loss spectroscopy was also successfully adapted to the needs of catalysis research and augmented by a unique combination with its surface-sensitive counterpart – soft X-ray absorption spectroscopy – which has been implemented as an in situ method at the BESSY synchrotron. Another instrument of strategic importance to the Department's research has been an ambient-pressure photoemission spectrometer (HP-XPS). A critical component of this instrument is a variable-energy "electron microscope" which transfers photoelectrons through a series of apertures into an electrostatic analyzer. The Department's HP-XPS has a long history, going back to the work of Jochen Block. Its present incarnation was built in collaboration with the group of Miquel Salmeron at the Advanced Light Source in Berkeley and became operational in 2007.

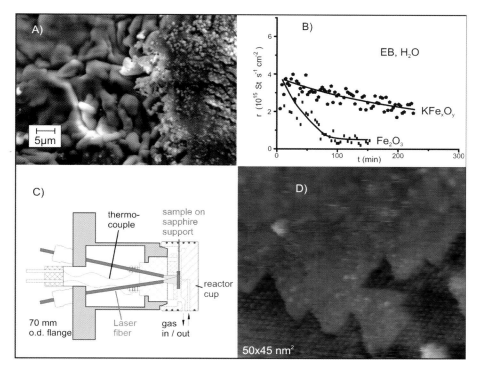

Fig. 6.6. *Potassium-promoted iron oxide for the dehydrogenation of ethylbenzene to styrene. (A) A SEM image of the technical catalyst with its internal interface. (B) Production rate of styrene over 1 cm² of model catalyst. (C) Micro-reactor for testing single-crystalline thin-film catalysts under ambient conditions with no background activity from the reactor itself. (D) STM image of the $K_2 Fe_{22} O_{34}$ active phase.*

Since no synthetic chemistry infrastructure was available at the Department (or, indeed, the Institute) before 2008, polycrystalline samples of catalysts had to be obtained from external, often industrial, partners. In order to produce model systems in house, researchers in the Department of Inorganic Chemistry developed a suite of instruments allowing the synthesis of metal oxides by physical vapor deposition of elements and by annealing procedures at ambient pressure. They chose the dehydrogenation of ethylbenzene to styrene on iron oxides as the subject of their first major study. Figure 6.6 summarizes the main results. The technical catalyst (A) is a complex convolution of phases, with the active sites located at the solid-solid interface. It was possible to synthesize well-ordered thin films (D) of the relevant ternary potassium iron oxide and to determine their chemical structure and reactivity. In parallel, Department members developed a micro-reactor device (B) allowing them to measure kinetic data (C) on such thin films. In this way, they were able to obtain experimental data needed for kinetic modeling under well-defined reaction conditions, which they could use to prove that the model reaction occurs in the same way as the reaction in the real-life system. Thin oxide

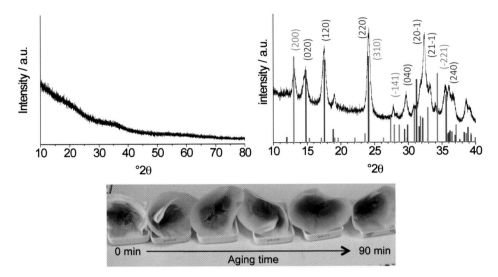

Fig. 6.7.   *Copper-based catalysts for methanol synthesis. A novel device for controlled precip-*
itation enabled separation of blue from green products. Structural analysis (top left)
revealed that the blue products are disordered nanocrystalline materials furnishing
poor catalysts. The green products are mixtures of two phases, malachite (violet) and
auricalcite (red). By systematically optimizing the reaction conditions it was possible to
prepare phase-pure green products and thereby to improve the synthesis of the working
catalyst based on pure malachite precursors. In the X-ray diffraction pattern (top right),
the features are labeled by the Miller indices, indicating the diffraction lattice plane of
the crystal; ° 2θ is the diffraction angle.

film research moved to the newly established Department of Chemical Physics in
1998, but a strong synthesis group remains within the Department of Inorganic
Chemistry.

This work enabled Department members to understand and control the details
of catalyst synthesis with a new level of precision. This improved control of materials quality and catalytic performance could then be applied to the preparation
of commercially relevant synthetic catalysts. Figure 6.7 shows as an example the
preparation of a Cu-based catalyst used for methanol synthesis. After constructing
a new device for computer-controlled synthesis group, its members were able to
resolve the precipitation process of the solid into two phases, as can be discerned
from the colors of the respective solids and the corresponding X-ray diffraction
patters. Time-resolved studies allowed them to answer a long-standing question
regarding the best precursor phase for the active catalyst – in favor of the zincian
malachite phase as opposed to the auricalcite phase; earlier attempts by a global
industry consortium to settle the question of the best precursor phase had been
unsuccessful. They were also able to explain the origin of the "chemical memory
effect," by recognizing that the non-translational structure of the Cu metal catalyst is controlled by the kinetics of the decomposition of the colored precursors

shown in Fig. 6.7. By controlling the defect structure of the Cu nanoparticles, they were further able to demonstrate the relevance of precursor synthesis to the final performance of the catalyst.

Several complementary in situ methods and electron microscopy were employed in this effort. Its successful completion required the targeted application of a broad portfolio of experimental techniques and is a testament to the highly effective collaboration within the Department. The Department comprises six Research Groups, each of which covers a certain field of competence (see below). Group leaders are responsible for teams of post-docs, students and engineering support, as well as for the transfer of know-how from one generation of collaborators to the next. Research, however, is commonly carried out in Project Teams, consisting of members of multiple Research Groups who collaborate to advance a specific topic of research, moving from a phenomenological to an analytical understanding of the subject. This bi-level collaborative structure has been in place since 2005.

As mentioned earlier, visualization of the non-translational structure of active catalysts has turned out to be of enormous value in finding strategies for analyzing quantitative data, and electron microscopy has provided particularly perspicacious visualizations. The Department's microscopes are stabilized by large masses of concrete in good contact with the native ground, which has turned out to be at least as good a method as the earlier, more complex approach based on active stabilization of platforms suspended in double-walled towers.

Figure 6.8 shows typical electron microscope images of catalysts. They demonstrate the need to understand structure on different length scales when analyzing the functioning of catalysts. The scanning electron microscope image shows a mesoporous silicate structure with hexagonal channels and their regular congestions. These unwanted long-range structures arise from the dynamics of the templating micelles. It is not possible to use transmission electron microscopy (TEM) to verify the homogeneity of the material, as the long-range channel modulation destroys the resolution of the TEM image by projecting the variable pore diameter onto the image plane. Whereas in the atomic domain aberration-corrected TEM is often an effective means to observe the dynamics of small objects, such as rafts of gold atoms anchored on purposely modified carbon nanotubes, which form highly effective catalysts for oxidations of biomass. The internal structure of the raft is fully dynamical at an observation time interval of 20 s, but the raft as a whole is firmly bound to its support and can even survive reactions in water and oxygen.

By following a research trajectory from overwhelmingly complex, high-performing catalysts to simplified model systems capable of maintaining a catalytic function of interest, work in the Department has shown repeatedly that complex solid structures are required to stabilize active phases, but that active phases are often chemically simple. In several cases, these active phases could be matched by model systems obtained through exacting physical preparation. The active species could then be identified and a theoretical description of the catalytic process in question provided whereby, in these instances, the gap between surface science and high-performance catalysis could be closed.

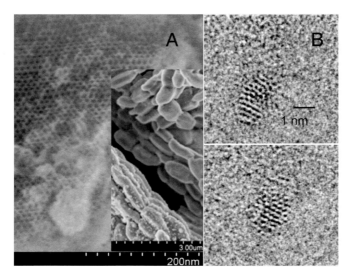

Fig. 6.8.   *Electron microscopy in catalysis. Plate (A) shows a mesoporous silicate SBA 15 sample at different magnifications. The hexagonal mesopores of 7 nm average diameter exhibit fluctuations with congestions every few hundreds of nanometers. Plate (B) shows a raft of gold atoms supported by carbon nanotubes. The top and bottom images attest to a fluctuation of the internal structure, observed over an interval of 20 s (given by the image recording time). The same atoms forming a three-dimensional nanoparticle appear static when observed under the same conditions.*

The Director of the Department of Inorganic Chemistry is Robert Schlögl, born in 1954 in Munich. He studied chemistry at the Ludwig Maximilians University in Munich, where he also completed his PhD on graphite intercalation compounds in 1982 under Gerhard Ertl. Prior to his appointment as Director at the FHI in 1994, he held a Professorship in Inorganic Chemistry at the University of Frankfurt (1989–1994).

The following researchers previously or currently affiliated with the Department of Inorganic Chemistry served as research group leaders:

- In situ Diffraction and Synthesis, led by Malte Behrens (PhD in Chemistry 2006, Christian-Albrechts-Universität Kiel, advisor Wolfgang Bensch; at FHI since 2006)

- Geometric Structure, led by Josef Find (PhD in Chemistry in 1998, Technische Universität Berlin, supervisor Robert Schlögl; at FHI from 1994 until 1999)

- Geometric Structure, led by Daniel Herein (PhD in Chemistry in 1995, Johann Wolfgang Goethe-Universität, supervisor Robert Schlögl; at FHI from 1994 until 1998)

- Emmy Noether Research Group, led by Christian Hess (PhD in Physical Chemistry in 2001, Freie Universität Berlin, supervisor Gerhard Ertl; at FHI from 1998 until 2008)

- Emmy Noether Research Group on High Temperature Catalysis, led by Raimund Horn (PhD in Chemistry 2003, Technische Universität Berlin, advisor Robert Schlögl; at FHI since 2007)

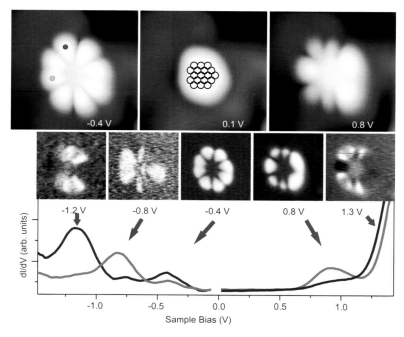

Fig. 6.10. *Set of STM images of flat Au$_{18}^{4-}$ cluster at three different tunneling voltages (upper panels) along with scanning tunneling spectra (lower panel) measured from −2.0 eV to +2.0 eV at the positions marked by the colored dots in the image taken at ∼ −0.4 eV (leftmost upper panel). The current images (middle panel) have been obtained at voltages corresponding to the observed maxima of the scanning tunneling spectra.*

the dots in the leftmost image of the upper panel. When the tip is placed in a position where the electron density vanishes, the STS peak disappears. The five so-called current images (middle panel) have been taken at voltages corresponding to the respective peak positions (or electronic energies). Note that the 18th Au atom (i.e., the rightmost one in the honeycomb-like structure shown in the center upper panel) makes the geometric structure of the Au$_{18}^{4-}$ cluster asymmetric, which is clearly reflected in the asymmetry of the electron densities. If this extra Au atom were absent, the electron densities would be symmetric. Hence the measured electronic structure disclosed details of the geometric structure.

Apart from the choice of a suitable oxide/metal system, members of the Department have also been able to control the system's properties by making use of the so-called strong metal support interaction (SMSI), in which an oxide film is grown over supported metal particles, a particularly feasible feat when reducible oxides are used as supports. The third case study that we will recount relied on the SMSI of platinum (Pt) nanoparticles with a reducible support provided by a Fe$_3$O$_4$ film grown on a Pt single crystal. After heating to 850 K, the capacity of the system to adsorb carbon monoxide (CO) is drastically reduced, which is typical for the SMSI effect. A close look at the surface images obtained by STM revealed

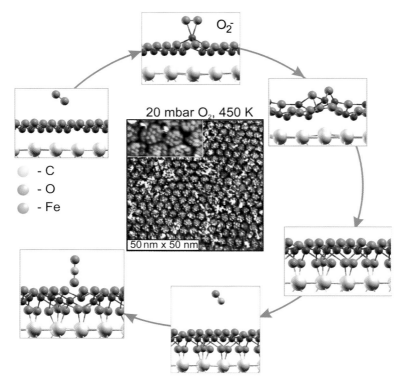

Fig. 6.11.   *Acts in forming the active tri-layer phase starting from FeO(111)/Pt(111) and its reaction with CO to form CO₂ based on density functional theory calculations. At center is an STM image of tri-layer FeO₂/Pt(111) formed at 20 mbar of O₂ at 450 K; the inset shows an atomic-resolution image.*

well-structured and facetted nanoparticles. However, images with atomic resolution showed a corrugation that did not stem from platinum but rather from a well-known, well-ordered double-layer film of ferrous oxide (FeO). One may thus reduce the complexity of the model system to a bilayer FeO film on a Pt single crystal, whose structure has been studied in detail and characterized at the atomic level. The 10 % mismatch between the FeO lattice constant and that of Pt gives rise to a characteristic Moiré pattern in the STM image. The bi-layer film is nonreactive under ultra-high vacuum conditions. However, the situation changed dramatically with respect to CO oxidation at ambient conditions (1 atm) in a reactor affording a careful control of the relative amounts of oxygen (one part, 20 mbar), carbon monoxide (two parts, 40 mbar) and helium as a buffer gas. Ramping up the temperature linearly at a rate of 1 K per minute from 300 K to 455 K, the CO oxidation ignited at 430 K.

The intriguing finding was that the catalytic activity of the FeO/Pt system in question was more than an order of magnitude greater than that of clean platinum (at 450 K). Usually SMSI diminishes catalytic activity; whereas, here researchers in the Department observed a strong *enhancement*. Further detailed experimental and theoretical (density functional theory-based) investigations revealed a scenario that made sense of this enhancement. The scenario is presented in Fig. 6.11. In Act 1, oxygen interacts with the bi-layer FeO film on Pt by pulling up an iron atom above the adsorbed-oxygen layer. This lowers the energy required to remove an electron at the interface to allow for an electron transfer to oxygen. This results in the formation of a transient O_2^{2-} molecule (Act 2), which dissociates and, at a higher oxygen coverage, gives rise to a local O-Fe-O tri-layer (Acts 3 and 4). The central panel of Fig. 6.11 displays an STM image of such a tri-layer formed in situ at an elevated O_2 pressure in the microscope. It owes its appearance mainly to the Moiré structure of the FeO bi-layer and covers about 80–90% of the surface. When exposed to CO (Act 5), the tri-layer oxidizes it to CO_2 (Act 6), leaving behind an oxygen vacancy in the film, which fills in again if the oxygen pressure is sufficiently high. Thereby, the trilayer is restored. If, however, the gas phase is oxygen-poor, the tri-layer is consumed and the reaction stops. Further investigations in the Department have confirmed that the iron oxide film de-wets the Pt single crystal surface under oxygen-poor reaction conditions by forming small iron oxide particles, which leaves the Pt crystal surface bare. The bare Pt surface then determines the reactivity of the system. Heating of the de-wetted surface in vacuum leads to the formation of the FeO bi-layer again which, at higher oxygen pressure, transforms into the tri-layer.

Freund and coworkers found as a corollary that a properly designed oxide film on a metal support can promote electron transfer to an adsorbed molecule. One such combination of materials is a bi-layer of MgO on Ag, which leads to the formation of a stable O_2^- molecular ion even under ultra-high vacuum conditions. A key experimental advance that made establishing this scenario possible was the development of an ultra-high vacuum electron spin resonance (ESR) spectrometer capable of determining magnetic properties of para- and ferro-magnetic species on surfaces, including their alignment and orientation. Hence they were able to conclude that the electron transfer to oxygen is the key step in initiating the oxidation reaction. This is in accord with Mott & Cabrera's theory of oxidation mentioned above and the long-neglected theories of catalytic activity proposed by Georg-Maria Schwab and Feodor Feodorovich Volkenshtein in the 1950s and 1960s. It has been the availability of 21st century techniques and apparatus to study surface interactions on an atomic level that brought us back to the future.

Advancing such innovative, enabling instrumentation has been among the Department's core activities. Other instruments developed by the Department include: a photoemission electron microscope (PEEM) with ultimate resolution attained by implementing corrections for both chromatic and spherical aberrations; a micro calorimeter whose sensitivity is sufficient to measure temperature-dependent heats of adsorption on nano-particles with aggregate sizes down to about a hundred atoms; and a photon STM, which adds chemical sensitivity through local excitation of a fluorescence signal by electrons from the tip.

The Director of the Department of Chemical Physics is Hans Joachim Freund, born in 1951 in Solingen. He studied Physics and Chemistry at the University of Cologne. As a graduate student in Georg Hohlneicher's group, he set up, in 1973, an X-ray photoelectron spectrometer (XPS) to study multielectron excitations of gaseous and adsorbed species. Prior to his appointment as Director at the FHI in 1996, he held Professorships at the Universities of Erlangen-Nürnberg (1983–1987) and Bochum (1987–1996), where he launched his research on model systems in catalysis.

The following researchers previously or currently affiliated with the Department of Chemical Physics for periods of several years had a noticeable impact on the Departments's work:

- Katharina Al-Shamery (PhD in Chemistry 1989, Universität Göttingen, adviser Prof. M. Quack; FHI 1996–1999; at present Professor at Carl von Ossietzky-Universität, Oldenburg, Germany)

- Marcus Bäumer (PhD in Chemistry 1994, Ruhr-Universität Bochum, adviser Klaus Rademann; FHI 1996–2002; at present Professor at Universität Bremen, Germany)

- Aidan Doyle (PhD in Chemistry 2000, University of Limerick, FHI 2002–2004; at present Associate Professor at Manchester Metropolitan University, UK)

- Javier Giorgi (PhD in Chemistry 1999, University of Toronto; FHI 2000–2002; at present Associate Professor at University of Ottawa, Canada)

- Markus Heyde (PhD in Chemistry 2001, Humboldt Universität zu Berlin, adviser Klaus Rademann; at FHI since 2003)

- Thorsten Klüner (PhD in Theoretical Chemistry 1997; Ruhr-Universität Bochum; adviser Hans-Joachim Freund; FHI 1994–2004; at present Professor at Carl von Ossietzky-Universität, Oldenburg, Germany)

- Christiane Koch (PhD in Theoretical Physics 2002, Humboldt Universität zu Berlin; adviser Hans-Joachim Freund; FHI 1998–2003; at present Professor Universität Kassel, Theoretische Physik III, Kassel)

- Helmut Kuhlenbeck (PhD in Physics 1988, Universtät Osnabrück, adviser Manfred Neumann; at FHI since 1996)

- Jörg Libuda (PhD in Physical Chemistry 1996; Ruhr-Universität Bochum; adviser Hans-Joachim Freund; FHI 1993–2005; at present Professor at Universität Erlangen-Nürnberg, Germany)

- Randall Meyer (PhD in Chemistry 2001, University of Texas at Austin, FHI 2001–2004; at present Assistant Professor at University of Illinois at Chicago, USA)

- Niklas Nilius (PhD in Physics 2001, Humboldt Universität zu Berlin; adviser Hans-Joachim Freund; at FHI since 2003)

- Zhihui Qin (PhD in Physics 2006;ÊInstitute of Physics, Chinese Academy of Sciences; FHI 2006–2009; at present Associate Professor at Wuhan Institute of Physics and Mathematics, Chinese Academy of Sciences, China)

- Thomas Risse (PhD in Chemistry 1996, Ruhr-Universtät Bochum; adviser Hans-Joachim Freund; FHI 1997–2010; at present Professor at Freie Universität, Berlin, Germany)

- Günther Rupprechter (PhD in Physical Chemistry 1992; Leopold Franzens Universität Innsbruck, Austria, adviser Prof. K. Hayek, FHI 1998–2006; at present Professor at Technische Universität Wien, Austria)

- Swetlana Schauermann (PhD in Chemistry 2005, Humboldt Universität zu Berlin; adviser Hans-Joachim Freund; at FHI since 2005)

- Thomas Schmidt (PhD in Physics 1994, Universität Hannover, adviser Martin Henzler; at FHI since 2008)

- Shamil Shaikhutdinov (PhD in Physics 1986, Moscow Institute of Physics and Technology, adviser Eduard Michailovich Trukhan; at FHI since 1999)

- Dario Stacchiola (PhD in Physical Chemistry 2002, University of Milwaukee, FHI 2005–2007; at present Assistant Professor at Michigan Technological University, Houghton, USA)

- Martin Sterrer (PhD in Chemistry 2003, Universität Wien, adviser Erich Knözinger; at FHI since 2003)

- Kazuo Watanabe (PhD in Chemistry 1998; The Graduate University for Advanced Studies, Tokyo, Japan; FHI 2004–2009; at present Associate Professor at Tokyo University of Science; Japan)

Department of Molecular Physics

Macroscopic reaction rates as observed, e.g., in the experiments of Michael Polanyi and others during the 1920s and 30s, represented averages over zillions of elementary collisions, whose identity and nature remained largely unknown – as did their relation to the molecular forces involved. This situation has been greatly remedied through the use of molecular beams whose deployment has made it possible to break with the bulk past and launch a new era in reaction kinetics based on the direct study of the dynamics of the underlying elementary collisions. Although the transition to the chemical/molecular dynamics era would materialize fully only three decades later and on the American continent, molecular beam methods have their roots in Europe, originating in part at Haber's institute. In 1921, Hartmut Kallmann and Fritz Reiche proposed a molecular beam experiment designed to find out whether individual polar molecules – as opposed to polar molecules in the bulk – carry an electric dipole moment. A beam of polar molecules was to be sent through an inhomogeneous electric field and its deflection monitored. Kallmann & Reiche presumed that, while the beam's dilution would preclude any bulk interaction among the molecules, the directionality of the molecules in the beam would make their deflection, if any, measurable. Kallmann & Reiche thereby tapped into a key feature of the molecular beam method, as later characterized by Otto Stern, who extolled the method's "simplicity and directness," emphasizing that it "enables us to make measurements on isolated neutral atoms or molecules with macroscopic tools…[and thereby] is especially valuable for testing and demonstrating directly fundamental assumptions of theory."

Kallmann & Reiche's paper prompted Stern to publish his proposal for what was to become the Stern-Gerlach experiment to test whether space quantization was real. Its demonstration, carried out in Frankfurt in 1922 by Stern and Walther Gerlach, ranks among the dozen or so canonical experiments that ushered in the heroic age of quantum physics. Perhaps no other experiment is so often cited for elegant conceptual simplicity. Among the descendants of the Stern-Gerlach experiment and its key concept of sorting quantum states via space quantization are the prototypes for nuclear magnetic resonance, optical pumping, the laser and atomic clocks, as well as incisive discoveries such as the Lamb shift and the anomalous increment in the magnetic moment of the electron, which launched quantum electrodynamics.

Despite their import, impact and lineage, neutral molecular beams and the gas-phase chemistry and physics that come along with them, had been absent at the institute from about 1933 until 2002, when Gerard Meijer was appointed to establish a department for the "pursuit of molecular physics and spectroscopy." The Department of Molecular Physics that ensued launched a research program organized around translationally cold neutral molecules and molecular beam spectroscopy of clusters and biomolecules, both neutral and ionized. The former topic in particular, which involves deceleration and trapping of molecules by means of electric and magnetic fields, can be regarded as an extrapolation of what Polanyi, Kallmann, Reiche and others had sought to do in their time.

In multipole focusers as well as in the deflection elements used in typical molecular beam experiments, the field gradient is perpendicular to the beam axis and exerts no force parallel to the beam axis. The forward (longitudinal) velocity distribution of a supersonic molecular beam is centered at a rather high velocity, ranging typically between 300 and 2000 m/s, depending on the molecular mass and the source conditions. Even at the low end of this velocity range, the kinetic energy of the molecules is on the order of 100 K, when expressed as energy divided by Boltzmann's constant. This is much larger than the depth of any potential energy well that can be realized for typical polar molecules using an electrostatic field, which amounts to only about 1 K. Therefore, a direct longitudinal confinement of the beam molecules by an electrostatic field is impossible.

However, the molecules' longitudinal and transversal velocity spreads are about the same, corresponding to a kinetic energy of about 1 K. Therefore, a potential well – with a gradient parallel to the molecular beam axis – could longitudinally confine the molecules, provided the potential moved along with the molecules in the beam at their most probable velocity. The confinement would be just as effective as the usual transversal confinement by a multipole focuser mounted parallel to the beam. Thus, a multipole focuser mounted perpendicular to the beam would produce the requisite longitudinal confinement, forcing the molecules on the beam axis to oscillate, both in position and velocity, around the center of the perpendicular multipole. Although the most probable velocity of the beam molecules would not be affected, the pack of molecules would remain flocked together while moving in the forward direction – it would be longitudinally focused (bunched). Moreover, if the velocity of this moving potential energy well were variable, a fraction of the beam molecules could be brought to any desired final velocity. So for instance, in order to decelerate the beam molecules, the potential well would have to be gradually slowed down such that the molecules in the beam would spend more time on the leading slope of the longitudinal potential well, thereby feeling a force opposing their motion. And vice versa, in order to accelerate the molecules, the potential well would have to be gradually sped up, thus pushing the molecules forward on the trailing slope of the potential well. The hypothetical apparatus just described functions almost exactly like a real-life Stark decelerator (or accelerator). However, rather than moving the electrodes that generate the longitudinal confinement, the elements of a static electrode array are energized or grounded synchronously with the movement of the molecules, thereby generating a traveling potential energy well. Depending on the timing sequence used to energize or ground the electrodes, the molecules can be either transported along the beam axis at a constant velocity or gradually decelerated or accelerated to any desired final velocity.

Stark manipulation of molecular beams has been considered and tried before. Electric field deceleration of neutral molecules was first attempted in 1959 at MIT by John King, who strived to produce a slow ammonia beam intended for a maser with an ultra-narrow linewidth. In the 1960s, at the University of Chicago, Lennard Wharton constructed an eleven meter long molecular beam machine for the acceleration of ground-state LiF molecules from 0.2 to 2.0 eV, with the goal of studying reactive scattering at hyperthermal collision energies. Both of these experiments were unsuccessful and were abandoned after the graduation of the PhD students involved. The first successful experimental demonstration of Stark deceleration took place in 1999 in Meijer's laboratory at the University of Nijmegen, where a beam of metastable CO molecules was slowed down from 225 m/s to 98 m/s. It was also Meijer who coined the term "Stark decelerator" or "Stark accelerator," for Johannes Stark, who prior to his embroilment in "German Physics," discovered the eponymous effect, by which an electric field characteristically shifts the energy levels of atoms and molecules.

Interest in Stark manipulation of molecular beams had been rekindled in the 1990s by the growth of "cold molecules" as a research topic in atomic, molecular and optical physics. Indeed, Stark deceleration has largely shaped the field of cold molecule research, as it became almost instantly the "workhorse" of the field. Moreover, the quantum-state selected molecular beam that exits a Stark decelerator has a tunable velocity, which is ideally suited for many applications. For instance, decelerated beams can be used in high-resolution spectroscopic studies to extend the available observation time and, by virtue of the uncertainty principle, to improve the attainable energy resolution. Decelerated beams also enable the study of (in)elastic collisions and reactive scattering down to zero collision energy and the study of the threshold behavior involved. Last but not least, a Stark decelerator enables trapping of neutral polar molecules.

Traps are key to further research in the field of cold molecules, in which the production and study of quantum degenerate gases of polar molecules is a particularly prominent objective. Trapping can also be used to prolong observation times to such an extent that radiative lifetimes of metastable molecular states can be accurately measured or the effects of the black-body radiation on molecules investigated. As John Fenn put it, "[b]orn in leaks, the original sin of vacuum technology, molecular beams are collimated wisps of molecules traversing the chambered void that is their theatre ... On stage for only milliseconds between their entrances and exits, they have captivated an ever growing audience by the variety and range of their repertoire." That the time the molecules are on stage is normally limited to milliseconds follows from the typical speed of the molecules (hundreds of meters per second) and the length of the vacuum chamber (a meter).

This time limitation can in part be overcome with a storage ring, a type of electrostatic trap in which low-field seeking beam molecules travel inside a one-meter vacuum chamber for over a mile, thus stretching the duration of their performance not by extending the theatre but by making the best of it. In a storage ring the molecules are kept in orbit by an array of electrostatic focusing elements. The most straightforward way of obtaining a storage ring for neutral molecules is to bend a single hexapole focuser onto itself to form a torus. Such a toroidal ring, however, does not confine the molecules longitudinally; an injected packet of molecules would spread and eventually fill the entire ring uniformly. However, this problem can be overcome by breaking the symmetry of the ring, e.g., by cutting it into segments separated by small gaps. The molecular beam packets can then be kept together (bunched) by changing the electric fields synchronously with their passage through the gaps, in analogy with the operation of a synchrotron for charged particles. The circling packets of molecules can then repeatedly interact, at well defined times and positions, with electromagnetic fields and/or with other atoms or molecules. When used as a low-energy collider to measure collision cross sections, the number of encounters per unit time scales as the square of the number of packets in the ring – which depends on the number of the ring's segments. Moreover, more segments enable storage of higher-density packets. Therefore, it's advantageous to cut the ring into as many segments as practically possible. A photograph of a molecular synchrotron consisting of 40 straight hexapoles, each 37 mm long, is shown in Fig. 6.12. Adjacent hexapoles, whose axes make an angle of $9°$ with respect to one another, are separated by gaps 2 mm wide. The resulting polygonal structure has a diameter of 0.5 m.

In one series of experiments performed in the Department, packets of Stark-decelerated deuterated ammonia ($^{14}ND_3$) molecules with a forward velocity of about 125 m/s were tangentially injected into the synchrotron. At its entry, a packet is several millimeters long and consists of about a million molecules. All these molecules are in the upper inversion doublet component of a single rotational level, the ground-state level of para-ammonia, hosted by the vibrational and

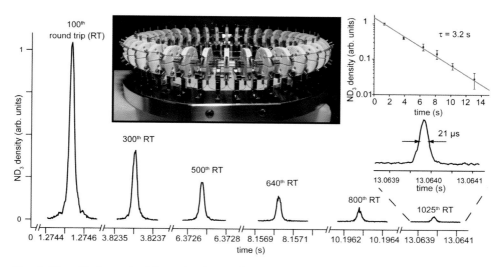

Fig. 6.12. *Photograph of the molecular synchrotron along with a plot of the measured density of ND_3 molecules as a function of time (in seconds) after loading for a selected number of laps (round trips, RT).*

electronic ground state. Once the molecular packet was inside the synchrotron, the hexapole fields were switched on, keeping the molecules both in orbit and transversely focused. By temporarily switching to a higher-voltage configuration whenever the packet passed through a gap, the molecules with a forward velocity spread of 1 m/s, corresponding to a temperature of 0.5 mK, were kept bunched while revolving.

Figure 6.12 shows the density of the ammonia molecules as a function of time after injection into the synchrotron. A total of 13 packets were injected at a rate of 10 Hz, after which the loading was suspended. The revolving packets then trailed each other by a distance of 3 hexapoles, with the first and the last injected packets 4 hexapoles apart. The molecules were laser-ionized and the ion signal due to the 13 packets was recorded for a selected number of laps. Even after 1025 laps, i.e. after the molecules traveled for more than a mile and passed a gap or hexapole 41,000 times, their signal could still be clearly discerned; the temporal width of 21 μs corresponds to a packet length of 2.6 mm. The density of the ammonia molecules in the synchrotron was seen to decrease exponentially with time at a rate of about 0.31 per second, which is caused, about equally, by collisions with background gas and excitation to an untrappable high-field seeking state by the blackbody radiation present in the room-temperature chamber. These measurements epitomize the level of control of molecular beams that can currently be achieved and set the stage for novel experiments that are yet to come.

Another research area currently pursued at the Department is spectroscopy of molecules, clusters, and biomolecules in the gas phase, one example being the infrared spectroscopy of gold clusters.

Conventional absorption spectroscopy is fairly difficult to apply, as the fragile clusters are highly dilute, i.e. in the form of molecular beams or confined in traps, where the attainable line-integrated absorber density is not sufficient for observing an absorption signal. An alternate route is to record the effect that the light exerts on the sample, in which case a sufficiently large fluence (number of photons/cm^2) may yield an observable signal, especially if the sample is photo-ionized by the incident radiation. The resulting ions or ionic fragments can then be mass-selectively detected with a unit efficiency. This so-called 'action spectroscopy' provides high sensitivity and is cluster-size selective. The crux of 'action spectroscopy' in the IR is a widely tunable laser of plentiful fluence, which is needed to induce the typically multiphoton processes involved.

During the last fifteen years, Meijer and coworkers have pioneered the application of infrared (IR) Free Electron Lasers (FELs) to obtain vibrational spectra of gas-phase species. The IR-FELs have provided access to weak modes in the far-infrared part of the spectrum corresponding to, e.g. metal-metal vibrations. Additionally, the spectroscopy in the gas phase has made it possible to extract the low-frequency modes which, in the case of deposited or embedded clusters, are often obscured by absorption in the substrate.

The absorption of far-IR photons can be detected by the 'messenger method,' which is based on forced evaporation of a weakly-bound ligand from the cluster complex upon the absorption of a small number of IR photons. The 'messenger method' assumes that only the metal cluster acts as a chromophore while the detached atomic or molecular ligand merely delivers the message about the absorption event without perturbing the structural properties of the cluster. This assumption is fulfilled for most transition metal clusters complexed with rare gas atoms. In combination with density functional theory calculations, the experimental IR (multi) photon dissociation, IR-(M)PD, spectra often enable an unambiguous determination of the clusters' structure.

Experimental information on the structure of charged gold clusters has been obtained from ion mobility measurements, trapped ion electron diffraction, and anion photoelectron spectroscopy. Significant structural differences between singly charged cationic and anionic gold clusters have been thereby identified and it could be concluded that the size at which the initially planar clusters morph into 3D structures strongly depends on their charge state. In combination with information available from vibrational spectroscopy of neutral gold clusters, researchers at the Department of Molecular Physics were able to obtain a complete picture of the charge state dependence of the structures. The charge-state dependence is exemplified in Fig. 6.13(a) for a gold cluster containing seven atoms. In this case, all three charge states (cation, neutral, anion) have distinct structures, although the rearrangement that takes place between the neutral species and the anion is only minor. At an increased electron density, the average coordination of the gold atoms decreases, resulting in the formation of increasingly open structures. For larger anionic gold clusters, such as Au_{19} and Au_{20}, tetrahedral structures have rather than electronic motion, thus conforming to the "Peierls type" mechanism. Furthermore, a detailed analysis of the transient changes of the electronic structure revealed oscillations arising from a vibrational mode in TbTe$_3$ (identified as the amplitude mode of the charge density wave), which could be attributed to the periodic lattice distortion of the CDW phase. The frequency of this mode, of about 2.5 THz, is consistent with the observed time delay of $\Delta t = 100\,fs$ required for the closing of the CDW gap.

The technique of time-, energy- and momentum-resolved photoemission spectroscopy has provided direct insights into the dynamics of the electronic structure of solids. In particular, the influence of electron-phonon coupling and other collective excitations on the (single-particle) band structure of solids could be observed directly through time-domain measurements. In future work, Wolf and coworkers will extend this technique by implementing new generation schemes for ultrashort VUV pulses, selective excitation of low energy modes and spin-resolved detection to obtain a complete picture of the electron dynamics in solids throughout the Brillouin zone.

The Director of the Department of Physical Chemistry is Martin Wolf, born in 1961 in Schwabach. He studied Physics at the Freie Universität Berlin and completed his PhD thesis at the FHI on surface photochemistry under Gerhard Ertl in 1991. In 1991–92 he was a Feodor Lynen Fellow at the University of Texas at Austin. Back at the FHI in 1992 as a staff scientist, he launched his research on ultrafast spectroscopy. Prior to his appointment in 2008 as a Director at the FHI, he held a Professorship in Physics at the Freie Universität Berlin (2000–2010).

The following research groups are currently supported by the Department of Physical Chemistry:

- Spatiotemporal Selforganization and Electrochemistry, led by Markus Eiswirth (PhD in Chemistry 1987, Ludwig-Maximilians-Universität München, adviser Gerhard Ertl; at FHI since 1990, previously Postdoctoral Fellow at Stanford University)

- Structural and Electronic Surface Dynamics, Max Planck Research Group led by Ralph Ernstorfer (PhD in Physics 2004, Freie Universität Berlin, adviser Frank Willig; at FHI since 2010; previously Postdoctoral Associate at the Max Planck Institut for Quantum Optics, Garching)

- Surface Femtochemistry and Ultrafast Carrier Dynamics, led by Christian Frischkorn (PhD in Physics 1997, Universität Göttingen, adviser Udo Buck; at FHI since 2008; previously Group Leader at the Freie Universität Berlin)

- Nanoscale Science, led by Leonhard Grill (PhD in Physics 200, Laboratorio TASC Trieste and University of Graz, adviser Silvio Modesti; at FHI since 2008; previously Research Group Leader at the Freie Universität Berlin)

- Time-resolved Second Harmonic Spectroscopy, led byAlexey Melnikov (PhD in Physics 1998, Moscow State University, adviser Oleg Aktsipetrov; at FHI since 2010; previously Group Leader at the Freie Universität Berlin)

- Complex Systems, led by Alexander Mikhailov (PhD in Physics 1976, Moscow State University, adviser Moisej Kaganov; at FHI since 1995; previously Leading Research Associate at the N.N. Semenenov Institut for Chemical Physics, Russian Academy of Sciences)

- THz Physics, led by Tobias Kampfrath (PhD in Physics 2006, Freie Universität Berlin, adviser Martin Wolf; at FHI since 2010; previously Postdoctoral Fellow at FOM Institute for Atomic and Molecular Physics, Amsterdam)

- Dynamics of Highly Correlated Materials, led by Patrick Kirchmann (PhD in Physics 2009, Freie Universität Berlin, adviser Martin Wolf; at FHI since 2011; previously Postdoctoral Fellow at Stanford University)

- Raman Spectroscopy, led by Bruno Pettinger (PhD in Physics 1972, FHI and Technische Universität München, adviser Heinz Gerischer, at FHI since 1970)

- Electron Dynamics, led by Julia Stähler (PhD in Physics 2007, Freie Universität Berlin, adviser Martin Wolf; at FHI since 2009; previously Postdoctoral Fellow at the University of Oxford)

Heller, Adam/Dieter Kolb/Krishnan Rajeshwar: The Life and Work of Heinz Gerischer, The Electrochemical Society Interface (Fall 2010), p. 37–40.

Hevesy, Georg von/Otto Stern: Fritz Habers Arbeiten auf dem Gebiete der physikalischen Chemie und Elektrochemie, Die Naturwissenschaften 16 (1928), p. 1062–1068.

[Hildebrandt, *Borrmann*] Hildebrandt, Gerhard: Gerhard Borrmann, Zeitschrift für Kristallographie 212 (1997), p. 618–626.

[Hildebrandt, *Laue*] Hildebrandt, Gerhard: Max von Laue, der "Ritter ohne Furcht und Tadel." In: [Hildebrandt, Treue, *Lebensbilder*] p. 223–244.

[Hildebrandt, Treue, *Lebensbilder*] Hildebrandt, Gerhard/Wilhelm Treue (Eds.): Berlinische Lebensbilder. Band 1: Naturwissenschaftler, (Einzelveröffentlichungen der Historischen Kommission zu Berlin 60), Berlin 1987.

[Hoffmann, *Havemann*] Hoffmann, Dieter/Dirk Draheim/Hartmut Hecht/Klaus Richter/Manfred Wilke: Robert Havemann. Dokumente eines Lebens, Berlin 1991.

[Hoffmann, *Havemann KWI*] Hoffmann, Dieter: Robert Havemann und das Kaiser-Wilhelm-Institut für physikalische Chemie (in preparation).

[Hoffmann, *Koppel*] Hoffmann, Dieter: Leopold Koppel (1854–1933). Bankier, Philantrop, Wissenschaftsmäzen, Berlin 2011.

[Hoffmann, *Laue*] Hoffmann, Dieter: „Nicht nur ein Kopf, sondern auch ein Kerl!" Zum Leben und Wirken Max von Laues (1879–1960). Physik Journal 9 (2010) p. 39–44.

[Hoffmann, Laitko, *Forschungshochschule*] Hoffmann, Dieter/Hubert Laitko: Die Deutsche Forschungshochschule, eine verhinderte institutionelle Innovation im Berlin der Nachkriegszeit (in preparation).

[Hirota, *Horiuti*] Hirota, Kozo: Juro Horiuti (1901–1979), in: Eley, D. D. (Ed.), Advances in Catalysis, New York 1981, p. xi–xiii.

[Horiuti, *Early Days*] Horiuti, Juro: Early Days in Electrochemistry, Journal of the Research Institute for Catalysis, Hokkaido University 22 (1975), p. 126–128.

Isaacson, Michael S.: Elmar Zeitler. The Chicago Years, Ultramicroscopy 49 (1993), p. 1–3.

[Jaenicke, *Kautsky*] Jaenicke, Lothar: Hans Kautsky (1891–1966) Erscheinungsbild und Bilderscheinung, Biospektrum 10 (2005), p. 532–535.

[Johnson, *Chemische Reichsanstalt*] Johnson, Jeffrey: Vom Plan einer Chemischen Reichsanstalt zum ersten Kaiser-Wilhelm-Institut: Emil Fischer, In: [Vierhaus, Brocke, *Forschung im Spannungsfeld*], p. 486–510.

[Johnson, *Chemists*] Johnson, Jeffrey: The Kaiser's Chemists. Science and Modernization in Imperial Germany, Chapel Hill 1990.

[Kalb, *Neumann*] Kalb, Stefanie: Wilhelm Neumann (1898–1965) – Leben und Werk unter besonderer Berücksichtigung seiner Rolle in der Kampfstoff-Forschung, Würzburg: Diss. 2005.

[Karrer, *Wieland*] Karrer, Paul: Heinrich Wieland. 1877–1957, Biographical Memoirs of Fellows of the Royal Society 4 (1958), p. 340–352.

[Klipping, Klipping, *Laboratory*] Klipping, Gustav/Ingrid Klipping: The Low-Temperatures Laboratory at the Fritz-Haber-Institute in Berlin, Indian Journal of Cryogenics 1 (1978), p. 254–265.

[Komarek, *Klipping*] Komarek, Peter/Klaus Lüders: Nachruf auf Gustav Klipping, Physik Journal 10 (2011), p. 45.

Lacmann, Rolf: Max Volmer und Iwan N. Stranski. In: [Hildebrandt, Treue, *Lebensbilder*] p. 329–342.

Lambert, Lotte/Thomas Mulvey: Ernst Ruska (1906–1988), Designer Extraordinaire of the Electron Microscope: A Memoir, Advances in Imaging and Electron Physics 95 (1996), p. 3–62.

[Leitner, *Immerwahr*] Leitner, Gerit von: Der Fall Clara Immerwahr. Leben für humane Wissenschaft, München 1993.

[Lemmerich, *Sturm*] Lemmerich, Jost: Aufrecht im Sturm der Zeit. Der Physiker James Franck 1882–1964, Diepholz 2007.

[L.F. Haber, *Poison*] Haber, Ludwig F.: The Poisonous Cloud. Chemical Warfare in the First World War, Oxford 1986.

[Lieb, *Kopfermann*] Lieb, Klaus-Peter: Theodor Schmidt and Hans Kopfermann: Pioneers in Hyperfine Physics, Hyperfine Interactions 136/137 (2001), p. 783–802.

[Löser, *Gründungsgeschichte*] Löser, Bettina: Zur Gründungsgeschichte und Entwicklung des Kaiser-Wilhelm-Instituts für Faserstoffchemie in Berlin-Dahlem (1914/19–1934), in: [Brocke, Laitko, *KWG Institute*] p. 275–302.

[Mehra, *Wigner*] Mehra, Jagdish (Ed.): The Collected Works of Eugene Paul Wigner, Part B: Historical, Philosophical, and Socio-Political Papers, Bd. 7, Berlin u.a. 2001.

[Meiser, *DFH*] Meiser, Inga: Die Deutsche Forschungshochschule (1947–1953), Diss. HU Berlin (in preparation).

[MPG, *FHI I*] MPG (Ed.): Fritz-Haber-Institut der Max-Planck-Gesellschaft, MPG Berichte und Mitteilungen 86/7 (1986).

[MPG, *FHI II*] MPG (Ed.): Fritz-Haber-Institut der Max-Planck-Gesellschaft, MPG Berichte und Mitteilungen 99/1 (1986).

[Niese, *Discovery*] Niese, Siegfried: The Discovery of Organic Solid and Liquid Scintillators by H. Kallmann and L. Herforth 50 Years ago, Journal of Radioanalytical and Nuclear Chemistry 241 (1999), p. 499–501.

[Nye, *Polanyi*] Nye, Mary Jo: Michael Polanyi and His Generation. Origins of the Social Construction of Science, Chicago 2011.

[Nye, *Tools*] Nye, Mary Jo: Working Tools for Theoretical Chemistry. Polanyi, Eyring and Debates Over the "Semiempirical" Method, Journal Computing Chemistry 28 (2007), p. 98–108.

[Pallo, *Farkas*] Pallo, Gabor: Prof. L. Farkas, 1904–1948. A Story of a Scientific Pioneer, Jerusalem 1998.

[Reinhardt, *BASF*] Reinhardt, Carsten: Über Wissenschaft und Wirtschaft. Fritz Habers Zusammenarbeit mit der BASF 1908 bis 1911. In: Albrecht, Helmuth (Ed.): Naturwissenschaft und Technik in der Geschichte. 25 Jahre Lehrstuhl für Geschichte der Naturwissenschaft und Technik am Historischen Institut der Universität Stuttgart, Stuttgart 1993, p. 287–315.

[Reitstötter, *Freundlich*] Reitstötter, Josef: Herbert Freundlich (1880–1941), Kolloid Zeitschrift 139 (1954), p. 1–11.

[Rossignol, *Ammoniak*] Rossignol, Robert Le: Zur Geschichte der Herstellung des synthetischen Ammoniaks, Die Naturwissenschaften 16 (1928), p. 1070–1071.

[Ruska, *Elektronenmikroskopie*] Ruska, Ernst: Die frühe Entwicklung der Elektronenlinsen und der Elektronenmikroskopie, Acta Historica Leopoldina 12 (1979), p. 3–136.

[Schäfer, *Reiche*] Schäfer, Clemens: Fritz Reiche 75 Jahre, Physikalische Blätter 14 (1958), p. 315–317.

[Schlüpmann, *Kopfermann*] Schlüpmann, Klaus: Vergangenheit im Blickfeld eines Physikers. Hans Kopfermann 1895–1963, 2002. http://www.aleph99.org/etusci/ks/index.htm

[Schmaltz, *Kampfstoff-Forschung*] Schmaltz, Florian: Kampfstoff-Forschung im Nationalsozialismus. Zur Kooperation von Kaiser-Wilhelm-Instituten, Militär und Industrie, (Geschichte der Kaiser-Wilhelm-Gesellschaft im Nationalsozialismus 11), Göttingen 2005.

[Schmaltz, *Thiessen*] Schmaltz, Florian: Peter Adolf Thiessen und Richard Kuhn und die Chemiewaffenforschung im NS-Regime, in: [Maier, *Gemeinschaftsforschung*] p. 305–351.

Schmidt-Ott, Dietrich: Die Deutsche Forschungshochschule in Berlin-Dahlem, Physikalische Blätter 8 (1952), p. 471–473.

Schmidt-Ott, Dietrich (und FHI Abteilungsleiter): Fritz-Haber-Institut der Max-Planck-Gesellschaft z.F.d.W. in Berlin-Dahlem, in: MPG (Ed.): Die Max-Planck-Gesellschaft und ihre Forschungsstellen, (Jahrbuch der Max-Planck-Gesellschaft zur Förderung der Wissenschaften e.V. 1961, Teil II), Göttingen 1962, p. 369–404.

[Schwankner, *Meergoldprojekt*] Schwankner, Robert: Das Meergoldprojekt. Berlin 1922–1927, Kultur und Technik 9 (1985/2), p. 65–85.

[Scott, *Polanyi*] Scott, William Taussig/Martin X. Moleski: Michael Polanyi. Scientist and Philosopher, New York 2005.

[Springer, *Kanig*] Springer, Jürgen: Nachruf auf Dr. rer. Nat. Gerhard Kanig, Colloid and Polymer Science 281 (2003), p. 1109–1110.

[Stern, *Freunde*] Stern, Fritz: Freunde im Widerspruch. Haber und Einstein, in: [Vierhaus, Brocke, *Forschung im Spannungsfeld*], p. 516–551.

[Stoltzenberg, *Haber*] Stoltzenberg, Dietrich J.: Fritz Haber. Chemiker, Nobelpreisträger, Deutscher, Jude. Eine Biographie, Weinheim 1998.

[Stoltzenberg, *KWIPhysChemie*] Stoltzenberg, Dietrich J.: Zur Geschichte des Kaiser-Wilhelm-Instituts für physikalische Chemie und Elektrochemie, Berichte zur Wissenschaftsgeschichte 14 (1991), p. 15–23.

[Szöllösi-Janze, *Haber*] Szöllösi-Janze, Margrit: Fritz Haber 1868–1934. Eine Biographie, München 1998.

[Terres, *Haber*] Terres, Ernst: Die Bedeutung Fritz Habers für die technische Chemie und die chemische Technik, Die Naturwissenschaften 16 (1928), p. 1068–1070.

[Thiessen, *Physikalische Chemie*] Thiessen, Peter Adolf: Die Physikalische Chemie im Nationalsozialistischen Staat, „Der deutsche Chemiker". Beilage zu Angewandter Chemie. Zeitschrift des Vereins Deutsche Chemiker Nr. 19, 1936. Reprinted in: Hentsche, Klaus/Ann Hentschel (Eds.): Physics and National Socialism. An Anthology of Primary Sources, Basel u.a. 1996, p. 134–137, Dokument 48.

[Thiessen, *Planck*], Thiessen, Peter Adolf: Max Planck prüft und verbessert, Annalen der Physik 500 (1988), p. 161–162.

[Weiss, *Spannung*] Weiss, Burghard: Höchste Spannung: Fritz Haber, Hartmut Kallmann und das "Tandem-Prinzip". Ein frühes Kapitel der Beschleuniger-Geschichte, Kultur & Technik 1 (1997), p. 42–49.

[Werner, *Haber Willstätter*] Werner, Petra/Angelika Irmscher (Eds.): Fritz Haber. Briefe an Richard Willstätter 1910–1934, (Studien und Quellen zur Geschichte der Chemie 6), Berlin 1995.

[Willstätter, *Haber*] Willstätter, Richard: Fritz Haber zum sechzigsten Geburtstag, Die Naturwissenschaften 16 (1928), p. 1053–1060.

[Wöhrle, *Haber Immerwahr*] Wöhrle, Dieter: Fritz Haber und Clara Immerwahr, Chemie in unserer Zeit 44 (2010), p. 30–39.

[Wolff*, Kallmann*] Wolff, Stefan: Hartmut Kallmann (1896–1978) – ein während des Nationalsozialismus verhinderter Emigrant verlässt Deutschland nach dem Krieg. In: Hoffmann, Dieter/Mark Walker (Eds.): „Fremde" Wissenschaftler im Dritten Reich, Göttingen 2011, p. 314–338.

[Wrangell, *Haber*] Wrangell, Margarethe von: Fritz Habers Bedeutung für die Landwirtschaft, Die Naturwissenschaften 16 (1928), p. 1071–1075.

Zeitler, Elmar: In Memory of Judith Reiffel, Ultramicroscopy 100 (2004) p. vii–ix.

[Zeitz, *Laue*] Zeitz, Katharina: Max von Laue (1879–1960). Seine Bedeutung für den Wiederaufbau der deutschen Wissenschaft nach dem Zweiten Weltkrieg, (Pallas Athene 16), Stuttgart 2006.

[Zott, *Haber*] Zott, Regine (Ed.): Fritz Haber in seiner Korrespondenz mit Wilhelm Ostwald sowie in Briefen an Svante Arrhenius, Berlin 1997.

References related to the Kaiser Wilhelm/Max Planck Society and to general institutional topics

[Andraschke, Hennig, *WeltWissen*] Andraschke, Udo/Jochen Hennig (Eds.): WeltWissen, 300 Jahre Wissenschaften in Berlin, München 2010.

Ash, Mitchell: Emigration und Wissenschaftswandel als Folgen der nationalsozialistischen Wissenschaftspolitik. In: Kaufmann, Doris (Ed.): Geschichte der Kaiser-Wilhelm-Gesellschaft im Nationalsozialismus. Bestandsaufnahme, Perspektiven und Forschung (Geschichte der Kaiser-Wilhelm-Gesellschaft im Nationalsozialismus 1/2). Göttingen 2000), S.610–631.

[Barkan, *Witches Sabbath*] Barkan, Diana Kormos: The Witches Sabath. The First International Solvay Congress in Physics, Science in Context 6 (1993), p. 59–82.

[Barkan, *Nernst*] Barkan, Diana: Walther Nernst and the Transition to Modern Physical Science. Cambridge 1999.

[Bartelt, *Berlin*] Bartelt, Hans-Georg: Aus der Geschichte der Physikalischen Chemie an der Friedrich-Wilhelms-Universität. Bunsen-Magazin 12(2010) p. 105–110

[Berlin, *Verordnungsblatt*] Verordnungsblatt der Stadt Berlin 1946, Nr. 20.

Beyler, Richard H.: "Reine" Wissenschaft und personelle "Säuberungen": Die Kaiser-Wilhelm-/Max-Planck-Gesellschaft 1933 und 1945, (Forschungsprogramm Geschichte der Kaiser-Wilhelm-Gesellschaft im Nationalsozialismus, Ergebnisse 16), Berlin 2004.

Biedermann, Wolfgang: Zur Finanzierung der Institute der Kaiser-Wilhelm-Gesellschaft zur Förderung der Wissenschaften Mitte der 20er bis Mitte der 40er Jahre des 20. Jahrhunderts. In: Parthey, Heinrich/Günther Spur (Eds.): Wissenschaft und Innovation (Gesellschaft für Wissenschaften, Wissenschaftsforschung Jahrbuch 2001), Berlin 2002, p. 143–172.

[Born, *Life*] Born, Max: *My Life: Recollections of a Nobel Laureate*, Scribner, New York, 1978.

[Brix, Ingwersen, Jaeschke, Repnow, *Beschleuniger*] Brix, P./H. Ingwersen/E. Jaeschke/R. Repnow: Linearbeschleuniger und Tandem-van-de-Graaffs. Werkzeuge der Schwerionenforschung, Die Naturwissenschaften 67 (1980), p. 265–273.

[Brocke, *System Althoff*] Brocke, Bernhard vom (Ed.): Wissenschaftsgeschichte und Wissenschaftspolitik im Industriezeitalter. Das „System Althoff" in historischer Perspektive, Hildesheim 1991.

[Brocke, Laitko, *KWG Institute*] Brocke, Bernhard vom/Hubert Laitko (Eds.): Die Kaiser-Wilhelm/Max-Planck-Gesellschaft und ihre Institute. Studien zu ihrer Geschichte: das Harnack-Prinzip, Berlin/New York 1996.

[Broser, *Geschichte*] Broser, Waldmar (Ed.): Chemie an der Freien Universität Berlin. Eine Dokumentation, (Wissenschaft und Stadt 8), Berlin 1988.

[Brown, Pike, *Optics*] Brown, Robert G.W./Edward Roy Pike: A History of Optical and Optoelectronic Physics in the Twentieth Century. In: Brown, Laurie M./ Abraham Pais/Sir Brian Pippard (Eds.): Twentieth Century Physics, Philadelphia 1995, p. 1385–1504.

[Burchardt, *Wissenschaftspolitik*] Burchardt, Lothar: Wissenschaftspolitik im Wilhelminischen Deutschland. Vorgeschichte, Gründung und Aufbau der Kaiser-Wilhelm-Gesellschaft zur Förderung der Wissenschaften, Göttingen 1975.

[Cahan, Meister] Cahan, David: An Institute for an Empire. Die Physikalisch-Technische Reichsanstalt 1871–1918, Cambridge University Press, 2004.

[Chayut, *Periphery*] Chayut, Michael: From the Periphery. The Genesis of Eugene P. Wigner's Application of Group Theory to Qauntum Mechanics, Foundations of Chemistry 3 (2001), p. 55–78.

[Czechowsky, Rüster, *Lindau*] Czechowsky, Peter/Rüdiger Rüster: 60 Jahre Forschung in Lindau – vom Fraunhofer-Institut zum Max-Planck-Institut für Sonnensystemforschung; eine Sammlung von Erinnerungen (1946–2006), Katlenburg-Lindau 2007.

[Daniels, *Radiation*] Daniels, Farington: The radiation hypothesis of chemical reaction. Chemical Reviews, vol. V (1928), p. 39–66.

[Deichmann, *Flüchten*] Deichmann, Ute: Flüchten, Mitmachen, Vergessen. Chemiker und Biochemiker in der NS-Zeit, Weinheim u.a. 2001.

[Deichmann, *Molecular*] Deichmann, Ute: "Molecular" versus "Colloidal". Controversies in Biology and Biochemistry, 1900–1940, Bulletin for the History of Chemistry 32 (2007), p. 105–118.

Deichmann, Ute: Proteinforschung an Kaiser-Wilhelm-Instituten von 1930 bis 1950 im internationalen Vergleich, (Geschichte der Kaiser-Wilhelm-Gesellschaft im Nationalsozialismus 21), Berlin 2004.

[Ede, *Rise*] Ede, Andrew: The Rise and Decline of Colloid Science in North America, 1900–1935, Burlington 2007.

[Einstein, *CPAE*] Einstein, Albert: Collected Papers of Albert Einstein, M.J. Klein, A.J. Kox, R.D. Schulmann (Eds.). Vol. 5. Princeton 1993.

[Engel, *Dahlem*] Engel, Michael: Geschichte Dahlems, Berlin 1984.

Flachowsky, Sören: Von der Notgemeinschaft zum Reichsforschungsrat. Wissenschaftspolitik im Kontext von Autarkie, Aufrüstung und Krieg, Stuttgart 2008.

[Fölsing, *Einstein*] Fölsing, Albrecht: Albert Einstein: A Biography. Penguin, 1998.

[Friese, *Japan*] Friese, Eberhard: Kontinuität und Wandel. Deutsch-japanische Kultur- und Wissenschaftsbeziehungen nach dem Ersten Weltkrieg. In: [Vierhaus, Brocke, *Forschung im Spannungsfeld*], p. 802–834.

[Gassert, Klimke, *1968*] Gassert, Philipp/Martin Klimke (Eds.): 1968. Memories and Legacies of a Global Revolt, Bulletin of the German Historical Institute, Supplement 6 (2009).

[Gavroglu, *London*] Gavroglu, Kostas: Fritz London: A Scientific Biography. Cambridge 1995.

[Gearhart, *Hydrogen*] Gearhart, Clayton: 'Astonishing Successes' and 'Bitter Disappointment'. The Specific Heat of Hydrogen in Quantum Theory, Archive for the History of Exact Sciences 64 (2010), p. 113–202.

Gerstengrabe, Sybille: Die erste Entlassungswelle von Hochschullehrern deutscher Hochschulen aufgrund des Gesetzes zur Wiederherstellung des Berufsbeamtentums vom 7.4.1933, Berichte zur Wissenschaftsgeschichte 17 (1994), p. 17–39.

Gerwin, Robert: 75 Jahre Max-Planck-Gesellschaft. Ein Kapitel deutscher Forschungsgeschichte Teile 1–3, Naturwissenschaftliche Rundschau 39 (1986), p. 1–10, p. 49–62, p. 97–109.

[Girnus, *Grundzüge*] Girnus, Wolfgang: Zu einigen Grundzügen der Herausbildung der physikalischen Chemie als Wissenschaftsdisziplin. In: Guntau, Martin/Hubert Laitko (Eds.): Der Ursprung der modernen Wissenschaften: Studien zur Entstehung wissenschaftlicher Disziplinen. Berlin 1987, p. 168–185.

[Groehler, *Tod*] Groehler, Olaf: Der lautlose Tod, Berlin 1978.

[Gruss, Rürup, *Denkorte*] Gruss, Peter/Reinhard Rürup/Susanne Kiewitz (Mitarb.): Denkorte. Max-Planck-Gesellschaft und Kaiser-Wilhelm-Gesellschaft. Brüche und Kontinuitäten 1911–2011, Dresden 2010.

[Hachtmann, *Wissensmanagement*] Hachtmann, Rüdiger: Wissensmanagement im „Dritten Reich". Geschichte der Generalverwaltung der Kaiser-Wilhelm-Gesellschaft (Geschichte der Kaiser-Wilhelm-Gesellschaft im Nationalsozialismus 15,1 und 15,2), Göttingen 2007.

[Hahn, *Leben*] Hahn, Otto, Mein Leben. München 1968.

[Hammerstein, *Wissenschaftspolitik*] Hammerstein, Notker: Die Deutsche Forschungsgemeinschaft in der Weimarer Republik und im Dritten Reich. Wissenschaftspolitik in Republik und Diktatur 1920–1945, München 1999.

[Hänseroth, Petschel, Pommerin, *175 Jahre*] Hänseroth, Thomas/Dorit Petschel/Reiner Pommerin: 175 Jahre TU Dresden. Die Professoren der TU Dresden 1828–2003, Köln 2003.

[Harnack, *Denkschrift*] Harnack, Karl Gustav Adolf von: Gedanken über die Notwendigkeit einer neuen Organisation zur Förderung der Wissenschaften in Deutschland, abgedruckt in: [MPG, *50 Jahre KWG/MPG*], p. 80–94.

[Hartley, *Report*] Hartley, Harold: Report on German Chemical Warfare. Organisation & Policy, 1914–1918. BNA, WO 243.

Heim, Susanne: Research for Autarky. The Contribution of Scientists to Nazi Rule in Germany (Geschichte der Kaiser-Wilhelm-Gesellschaft im Nationalsozialismus 4), Berlin 2001.

Heim, Susanne/Carola Sachse/Mark Walker (Eds.): The Kaiser Wilhelm Society under National Socialism, Cambridge 2009.

Henning, Eckart/Marion Kazemi: Dahlem – Domäne der Wissenschaft. Ein Spaziergang zu den Berliner Instituten der Kaiser-Wilhelm/Max-Planck-Gesellschaft im „deutschen Oxford", (Veröffentlichungen aus dem Archiv der Max-Planck-Gesellschaft 16/I), 4. Erw. und aktual. Aufl., Berlin 2009.

[Hermann, *Frühgeschichte*] Hermann, Armin: Frühgeschichte der Quantentheorie 1899–1913, Mosbach 1969.

[Hoffmann, *Einsteins Berlin*] Hoffmann, Dieter: Einsteins Berlin. Auf den Spuren eines Genies, Weinheim 2006.

[Hoffmann, *Planck*] Hoffmann, Dieter: Max Planck. Die Entstehung der modernen Physik. München 2008.

Hofmann, Heini: Geheimobjekt "Seewerk". Vom Geheimobjekt des Dritten Reiches zum wichtigsten Geheimobjekt des Warschauer Vertrages, 2. erweiterte Aufl., Zella-Mehlis/Meiningen 2008.

[Holton, *Pais Prize Lecture*] Holton, Gerald: Pais Prize Lecture. Of What Use is the History of Science? American Physical Society Forum on the History of Physics 2008, http://www.aps.org/units/fhp/newsletters/fall2008/pais.cfm.

[Jansen, Duncan, *Umdeutung*] Jansen, Michel/Tony Duncan: On the Verge of Umdeutung: John Van Vleck and the Correspondence Principle. I and II, Archive for the History of Exact Sciences 61 (2007), p. 553–624, 625–671.

[Jansen, *Schädlinge*] Jansen, Sarah: "Schädlinge." Geschichte eines wissenschaftlichen und politischen Konstrukts, 1840–1920. Frankfurt/M. 2003.

[Jenrich, *Ihne*] Jenrich, Franziska: Ernst von Ihnes Bauten für die Kaiser-Wilhelm-Gesellschaft in Berlin-Dahlem, Magisterarbeit FU Berlin, Berlin 2010.

[Johnson, MacLeod, *Disarmament*] Johnson, Jeffrey/Roy MacLeod: The War the Victors Lost. The Dilemmas of Chemical Disarmament, 1919–1926. In: Johnson, Jeffrey/Roy MacLeod (Eds.): Frontline and Factory. Comparative Perspectives on the Chemical Industry at War, 1914–1924, Dordrecht 2006, p. 221–245.

Klee, Ernst: Das Personenlexikon zum Dritten Reich. Wer war was vor und nach 1945, 3. Aufl., Frankfurt am Main 2011.

Kohl, Ulrike: Die Präsidenten der Kaiser-Wilhelm-Gesellschaft im Nationalsozialismus: Max Planck, Carl Bosch und Albert Vögler zwischen Wissenschaft und Macht, (Pallas Athene 5), Stuttgart 2002.

[Kohler, *Partners*] Kohler, Robert: Partners in Science. Foundations and Natural Scientists, Chicago 1991.

Krohn, Claus-Dieter/Patrick von zur Mühlen/Gerhard Paul/Lutz Winckler (Eds.): Handbuch der Emigration 1933–1945, Darmstadt 1998. Teil IV: Wissenschaftssemigration. p. 681–922.

[Laitko, *Innovationen*] Laitko, Hubert: Wissenschaft in Berlin – eine Problematik zwischen allgemeiner Geschichte und Wissenschaftsgeschichte, In: Berlingeschichte im Spiegel wissenschaftshistorischer Forschung – 300 Jahre Wissenschaft in Berlin. Materialien der wissenschaftlichen Konferenz vom 9. – 11. April 1987 anläßlich der 750-Jahr-Feier der Stadt Berlin. itw-kolloquien Nr. 64). Berlin 1987, S. 41 ff.

[Laitko, *Wissenschaft Berlin*] Laitko, Hubert *et al.*: Wissenschaft in Berlin. Von den Anfängen bis zum Neubeginn nach 1945, Berlin 1987.

[Landolt, *Antrittsrede*] Landolt, Hans: Antrittsrede, Sitzungsberichte der Preußischen Akademie der Wissenschaften 2 (1882), p. 723–723.

[Lepick, *Grande*] Lepick, Olivier: Le Grande Guerre Chemique, 1914–1918, Paris 1998.

Luxbacher, Günther: Roh- und Werkstoffe für die Autarkie. Textilforschung in der Kaiser-Wilhelm-Gesellschaft (Geschichte der Kaiser-Wilhelm-Gesellschaft im Nationalsozialismus 18), Berlin 2004.

Macrakis, Kristie: Surviving the Swastika. Scientific Research in Nazi Germany, New York 1993.

[Macrakis, *Rockefeller*] Macrakis, Kristie: The Rockefeller Foundation and German Physics under National Socialism, Minerva 27 (1989), p. 33–57.

[Maier, *Forschung*] Maier, Helmut: Forschung als Waffe. Rüstungsforschung in der Kaiser-Wilhelm-Gesellschaft und das Kaiser-Wilhelm-Institut für Metallforschung 1900 bis 1945/48, Göttingen 2007.

[Maier, *Gemeinschaftsforschung*] Maier, Helmut (Ed.): Gemeinschaftsforschung, Bevollmächtigte und der Wissenstransfer. Die Rolle der Kaiser-Wilhelm-Gesellschaft im System kriegsrelevanter Forschung des Nationalsozialismus,

(Geschichte der Kaiser-Wilhelm-Gesellschaft im Nationalsozialismus 17), Göttingen 2007.

[Maier, *Rüstungsforschung*] Maier, Helmut (Ed.): Rüstungsforschung im Nationalsozialismus. Organisation, Mobilisierung und Entgrenzung der Technikwissenschaften, (Geschichte der Kaiser-Wilhelm-Gesellschaft im Nationalsozialismus 3), Göttingen 2002.

[Marsch, *Notgemeinschaft*] Marsch, Ulrich: Notgemeinschaft der Deutschen Wissenschaft. Gründung und frühe Geschichte, 1920–1925, Frankfurt am Main u.a. 1994.

[Martinetz, *Gaskrieg*] Martinetz, Dieter: Der Gaskrieg 1914–1918. Entwicklung, Herstellung und Einsatz chemischer Kampfstoffe; das Zusammenwirken von militärischer Führung, Wissenschaft und Industrie, Bonn 1996.

Mehrtens, Herbert: Kollaborationsverhältnisse. Natur und Technikwissenschaften im NS-Staat und ihre Historie. In: [Meinel, Voswinkel, *MNT im NS*], p. 13–32.

[Meinel, Voswinkel, *MNT im NS*] Meinel, Christoph/Peter Voswinkel (Eds.): Medizin, Naturwissenschaft und Nationalsozialismus. Kontinuitäten und Diskontinuitäten, Stuttgart 1994.

[MPG, *Jahrbuch*] Max-Planck-Gesellschaft zur Förderung der Wissenschaften: Jahrbuch der Max-Planck-Gesellschaft zur Förderung der Wissenschaften, Göttingen/München 1951–2006.

[MPG, *MiMax*] Max-Planck-Gesellschaft zur Förderung der Wissenschaften (Ed.): Mitteilungen aus der Max-Planck-Gesellschaft zur Förderung der Wissenschaften, München 1952–1974.

[MPG, *50 Jahre KWG/MPG*] Generalverwaltung der Max-Planck-Gesellschaft (Ed.): 50 Jahre Kaiser-Wilhelm-Gesellschaft und Max-Planck-Gesellschaft zur Förderung der Wissenschaften 1911–1961, Beiträge und Dokumente, Göttingen 1961.

Müller, Falk: The Birth of a Modern Instrument and Its Development During World War II. Electron Microscopy in Germany from the 1930s to 1945. In: Hooijmaijers, Hans/Ad Maas: Scientific Research in WWII. What Scientists Did in the War, London u.a. 2009, p. 121–146.

[Nachmansohn, *Pioneers*] Nachmansohn, David: German-Jewish Pioneers in Science 1900–1933. Highlights in Atomic Physics, Chemistry and Biochemistry, Berlin 1979.

Nagel, Günter: Unterschätzte Rüstungsforschung. Erich Schumann und das II. Physikalische Institut der Universität. In: Karlsch, Rainer/Heiko Petermann (Eds.): Für und Wider "Hitlers Bombe". Studien zur Atomforschung in Deutschland, Münster 2007, p. 229–260.

[Palazzo, *Seeking*] Palazzo, Albert: Seeking Victory on the Western Front. The British Army and Chemical Warfare in WWI, Lincoln 2000.

Parthey, Heinrich: Bibliometrische Profile von Instituten der Kaiser-Wilhelm-Gesellschaft zur Förderung der Wissenschaften (1923–1943). Institute der

Chemisch-Physikalisch-Technischen und der Biologisch-Medizinischen Sektion (Veröffentlichungen aus dem Archiv zur Geschichte der Max-Planck-Gesellschaft 7), Berlin 1995.

[Planck, *Besuch*] Planck, Max: Mein Besuch bei Adolf Hitler, Physikalische Blätter (1947), p. 143.

[Podaný, *Heyrovský*] Podaný, Václav: Jaroslav Heyrovský (1890–1967) Chemiker und Johann Böhm (1895–1952) Chemiker. Schicksale zweier Wissenschaftler. In: Monika Glettler/Alena Míšková (Eds.): Prager Professoren 1938–1948. Zwischen Wissenschaft und Politik, (Veröffentlichungen zur Kultur und Geschichte im östlichen Europa 17), Essen 2001, p. 543–568.

[Rasch, *Kohlenforschung*] Rasch, Manfred: Geschichte des Kaiser-Wilhelm-Instituts für Kohlenforschung 1913–1943, Weinheim u.a. 1989.

Rasmussen, Nicolas: Picture Control. The Electron Microscope and the Transformation of Biology in America 1940–1960, Stanford 1997.

[Reinhardt, *Zentrale*] Reinhardt, Carsten: Zentrale einer Wissenschaft, In: Tenorth, Heinz-Elmar (Ed.): Geschichte der Universität Unter den Linden 1810–2010, Bd. 5: Transformation der Wissensordnung. Berlin 2010, p. 575–604.

[Renn, *Einstein Kontexte*] Renn, Jürgen/Giuseppe Castagnetti/Peter Damerow: Albert Einstein. Alte und neue Kontexte in Berlin. In: Kocka, Jürgen (Ed.): Die Königlich Preußische Akademie der Wissenschaften in Berlin im Kaiserreich, Berlin 1999, p. 333–354.

[Renn, Kant, *KWG/MPG*] Renn, Jürgen/Horst Kant: Kurze Geschichte der Kaiser-Wilhelm- und Max-Planck-Gesellschaft. Berlin 2011.

Renneberg, Monika/Mark Walker (Eds.): Science, Technology, and National Socialism, Cambridge 1994.

[Reulecke, *Leistungskampf*] Reulecke, Jürgen: Die Fahne mit dem goldenen Zahnrad. Der 'Leistungskampf deutscher Betriebe' 1937–1939. In: Peuckert, Detlev/Jürgen Reulecke (Eds.): Die Reihen fest geschlossen. Beiträge zur Geschichte des Alltags unterm Nationalsozialismus, Wuppertal 1981, p. 242–272.

[Rürup, Schüring, *Schicksale*] Rürup, Reinhard/Michael Schüring: Schicksale und Karrieren. Gedenkbuch für die von den Nationalsozialisten aus der Kaiser-Wilhelm-Gesellschaft vertriebenen Forscherinnen und Forscher, Göttingen 2008.

[Sauer, *Superconductivity*] Sauer, Tilman: Einstein and the Early Theory of Superconductivity, 1919–1922, Archive for the History of Exact Sciences 61 (2007), p. 159–211.

[Schüring, *Minerva*] Schüring, Michael: Minervas verstoßene Kinder. Vertriebene Wissenschaftler und die Vergangenheitspolitik der Max-Planck-Gesellschaft, (Geschichte der Kaiser-Wilhelm-Gesellschaft im Nationalsozialismus 13), Göttingen 2006.

[Schweer, *Stoltzenberg*] Schweer, Henning: Die Geschichte der Chemischen Fabrik Stoltzenberg bis zum Ende des zweiten Weltkrieges, Diepholz 2008.

[*Sehr visionär und kühn*] Sehr visionär und kühn. Rundtischgespräch anlässlich des 20. Jahrestags des Mauerfalls. Physik Journal 8 (2009), p. 24–28.

Resolved Photoemission from Oxygen Adsorbed on Nickel (100) in the p (2 × 2) Structure, Solid State Communications 25 (1978), p. 93–99.

[Franck, Einsporn, *Quicksilberdampfes*] Franck, James/Erich Einsporn: Über die Anregungspotentiale des Quecksilberdampfes, Zeitschrift für Physik 2 (1920), p. 18–29.

[Franck, Hertz, *Zusammenstöße*] Franck, James/Gustav Hertz: Über Zusammenstöße zwischen Elektronen und den Molekülen des Quecksilberdampfes und die Ionisierungsspannung desselben, Verhandlungen der Deutschen physikalischen Gesellschaft 16 (1914), p. 457–467.

[Franck, Reiche, *Helium*] Franck, James/Fritz Reiche: Über Helium und Parahelium, Zeitschrift für Physik 1 (1920), p. 154–160.

[Freundlich, *Chemistry*] Freundlich, Herbert: Colloid and Capillary Chemistry, New York 1926.

[Freundlich, Wreschner, *Elektrokapillarkurve*] Freundlich, Herbert/Marie Werschner: Über den Einfluß der Farbstoffe auf die Elektrokapillarkurve, Kolloid Zeitschrift 28 (1921), p. 250–253.

[Freundlich, *Kapillarchemie*] Freundlich, Herbert: Kapillarchemie. Eine Darstellung der Chemie der Kolloide und verwandter Gebiete, Leipzig 1922.

[Freundlich, *Kolloidchemie*] Freundlich, Herbert: Kapillarchemie und Kolloidchemie, Kolloid Zeitschrift 31 (1922), S.243–246.

[Freundlich, Nathansohn, *Lichtempfindlichkeit*] Freundlich, Herbert/Alexander Nathansohn: Über die Lichtempfindlichkeit des Arsentrisulfidsols, Kolloid Zeitschrift 28 (1921), p. 258–262.

[Freundlich, *Thixotropie*] Freundlich, Herbert: Über Thixotropie, Kolloidzeitschrift 46 (1928), p. 289–299.

[Freundlich, Rogowski, Söllner, *Ultraschallwellen*] Freundlich, Herbert/Friedrich Rogowski/Karl Söllner: Über die Wirkung der Ultraschallwellen auf thixotrope Gele, Zeitschrift für Physikalische Chemie 160 (1932), p. 469–472.

[Frey, *Kassette*] Frey, Friedrich: Verwendungsmöglichkeit und Konstruktion einer Kassette für Doppelaufnahmen mit dem Siemens-Übermikroskop in Normalausführung, Zeitschrift für technische Physik 23 (1942), p. 176–177.

[Fricke, Schoon, Schröder, *Umwandlungsreihe*] Fricke, Robert/Theodor Schoon/W. Schröder: Eine gleichzeitig röntgenographische und elektronenmikroskopische Verfolgung der thermischen Umwandlungsreihe γ-FeOOH–γ-Fe$_2$O$_3$–α-Fe$_3$O$_4$, Zeitschrift für Physikalische Chemie B 30 (1941), p. 13–22.

[Friedrich-Freksa, Kratky, Sekora, *Röntgeninterferenzen*] Friedrich-Freksa, Hans/ Otto Kratky/Aurelie Sekora: Auftreten von neuen Röntgeninterferenzen bei Einlagerung von Jod in Seidenfibroin vom Bombyxmori-Typ, Die Naturwissenschaften 32 (1944), p. 78.

[Frisch, Goodall, Greenhow, Holzwarth, Knight, *Single-Photon*] Frisch, Wolfgang/ David M. Goodall/Rodney C. Greenhow/Josef F. Holzwarth/Barry Knight: Single-Photon Infrared Photochemistry. Wavelength and Temperature Dependence of the Quantum Yield for the Laser-Induced Ionization of Water, in: Karl L. Kompa/Stanley Desmond Smith (Eds.): Laser-Induced Processes in

Molecules. Proceedings of the European Physics Society Division Conference at Heriot-Watt Univ., Edinburgh, Sept. 20–22, 1978 (Springer Proceedings in Physics 6), New York 1979, p. 283–285.

[Fujimoto, Kambe, Lehmpfuhl, Uchida, *Dunkelfeldtechnik*] Fujimoto, Fuminori/ Kyozaburo Kambe/Günter Lehmpfuhl/Yuji Uchida: Dunkelfeldtechnik zur elektronenmikroskopischen Abbildung von Oberflächenstrukturen auf Einkristallen bei Energien zwischen 100 keV und 1000 keV, in: Gerhard Pfefferkorn (Ed.): Beiträge zur elektronenmikroskopischen Direktabbildung von Oberflächen, Bd. 8, Münster 1977, p. 417–425.

[Fujimoto, Kambe, Lehmpfuhl, *Electron Channeling*] Fujimoto, Fuminori/Kyozaburo Kambe/Günter Lehmpfuhl: Interpretation of Electron Channeling by the Dynamical Theory of Electron Diffraction, Zeitschrift für Naturforschung 29 A (1974), p. 1034–1044.

[Gerischer, *Halbleiter I*] Gerischer, Heinz: Über den Ablauf von Redoxreaktionen an Metallen und an Halbleitern. I. Allgemeines zum Elektronenübergang zwischen einem Festkörper und einem Redoxelektrolyten, Zeitschrift für Physikalische Chemie N. F. 26 (1960), p. 223–247.

[Gerischer, *Halbleiter II*] Gerischer, Heinz: Über den Ablauf von Redoxreaktionen an Metallen und an Halbleitern. II. Metall-Elektroden, Zeitschrift für Physikalische Chemie N. F. 26 (1960), p. 325–338.

[Gerischer, *Halbleiter III*] Gerischer, Heinz: Über den Ablauf von Redoxreaktionen an Metallen und an Halbleitern. III. Halbleiterelektroden, Zeitschrift für Physikalische Chemie N. F. 27 (1961), p. 48–79.

[Gerischer, Kolb, Schulze, *Optical Absorption*] Gerischer, Heinz/Dieter M. Kolb/ Wilfried Schulze: Optical Absorption Spectra of Matrix Isolated Silver Atoms and their Dependence on Matrix Properties, Journal of the Chemical Society, Faraday Transactions II 71 (1975), p. 1763–1771.

[Gerischer, Gobrecht, Kautek, *Semiconducting Materials*] Gerischer, Heinz/Jens Gobrecht/Wolfgang Kautek: The Applicability of Semiconducting Layered Materials for Electrochemical Solar Energy Conversion, Berichte der Bunsengesellschaft für physikalische Chemie 84 (1980), p. 2471–2478.

[Gimzewski, Sass, Schlittler, Schott, *Scanning Tunneling*] Gimzewski, James K./ Jürgen-Kurt Sass/Reto R. Schlittler, J. Schott: Enhanced Photon Emission in Scanning Tunneling Microscopy, Europhysics Letters 8 (1989), p. 435–440.

[Golze, Grunze, Hirschwald, Polak, *XPS-Study*] Golze, Manfred/Michael Grunze/ Wolfgang Hirschwald/M. Polak: A XPS-Study of the Intermediates in Nitrogen Dissociation on an Fe(111)-Surface, in: Gerhard Betz/Peter Braun (Eds.): Symposium on Surface Science (Obertraun/Austria 1983.01.31– 02.04),Contributions, Wien 1983, p. 186–191.

[Grund, Heide, *Biological Specimens*] Grund, Siegfried/Hans-Günther Heide: Wet Biological Specimens in the Electron Microscope. Transfer and Observation at Low Temperature, in: David J. Goodchild/V. Sanders (Eds.): Electron Microscopy 1974. Abstracts of Papers Presented to the Eighth International Congress on

Electron Microscopy. Held in Canberra, Australia. August 25–31 1974, Bd. 2. Canberra 1974, p. 46–47.

[Haber, Zisch, *Anregung*] Haber, Fritz/Walter Zisch: Anregung von Gasspektren durch chemische Reaktionen, Zeitschrift für Physik 9 (1922), p. 302–326.

[Haber, Jaenicke, Matthias, *Darstellung*] Haber, Fritz/Johannes Jaenicke/Friedrich Matthias: Über die angebliche Darstellung „künstlichen" Goldes aus Quecksilber, Zeitschrift für anorganische und allgemeine Chemie 153 (1926), p. 153–183.

[Haber, *Dispersion*] Haber, Fritz: Untersuchungen über die anomale Dispersion angeregter Gase, Die Naturwissenschaften 15 (1927), p. 174.

[Haber, *Gold*] Haber, Fritz: Das Gold im Meerwasser, Zeitschrift für angewandte Chemie 40 (1927), p. 303–314.

[Haber, *Körper*] Haber, Fritz: Über den festen Körper sowie über den Zusammenhang ultravioletter und ultraroter Eigenwellenlängen der Bildungswärme mit der Quantentheorie, Verhandlungen der Deutschen physikalischen Gesellschaft 13 (1911), p. 1117–1136.

[Haber, *Phasengrenzkräfte*] Haber, Fritz/Zygmunt Aleksander Klemensiewicz: Über elektrische Phasengrenzkräfte, Zeitschrift für Physikalische Chemie 67 (1909), p. 385–431.

[Haber, *Umwandelbarkeit*] Haber, Fritz: Über den Stand der Frage nach der Umwandelbarkeit der chemischen Elemente, Die Naturwissenschaften 14 (1926), p. 405–412.

[Haber, *Zeitalter*] Haber, Fritz: Das Zeitalter der Chemie, in: Fritz Haber: Fünf Vorträge aus den Jahren 1920–1923, Berlin 1924.

[Hamilton, *Neutron Diffraction*] Hamilton, Walter C./Armin Tippe: Neutron Diffraction Study of Decaborane, Inorganic Chemistry 8 (1969), p. 464–470.

[Haul, Scholz, *Grenzflächen-Reaktionen*] Haul, Robert/E. Scholz: Anodische Grenzflächen-Reaktionen an der Quecksilber-Tropfelektrode, Zeitschrift für Elektrochemie und angewandte physikalische Chemie 52 (1948), p. 226–234.

[Hauschild, Lüttringhaus, *Valenzwinkelstudien*] Lüttringhaus, Arthur/Kurt Hauschild: Valenzwinkelstudien, VI. Mitteilung: Zur Stabilität des Tetraederwinkels am Kohlenstoffatom, Berichte der Deutschen Chemischen Gesellschaft 73 (1940), p. 134–145.

[Helmcke, *Atlas*] Helmcke, Johann-Georg: Atlas des menschlichen Zahnes im elektronenmikroskopischen Bild, 2 Bde., Berlin 1953 und 1957.

[Helmcke, Krieger, *Diatomeenschalen*] Helmcke, Johann-Georg/Willi Krieger: Diatomeenschalen im elektronenmikroskopischen Bild, Teil 1–10, Berlin 1953–1977.

[Helmcke, Otto, *Konstruktionen*] Helmcke, Johann-Gerhard/Frei Otto: Lebende und technische Konstruktionen, Deutsche Bauzeitung 67 (1962), Heft 11, p. 885–886.

[Herrmann, *Bildverstärker*] Herrmann, Karl-Heinz: Bildverstärker in der Elektronenmikroskopie, in: Gottfried Möllenstedt: Höchstauflösung in der Elektronenmikroskopie, München 1973, p. 39–55.

[Herrmann, Kunath, Schiske, Weiss, Zemlin, *Coma-Free Alignment*] Herrmann, Karl-Heinz/Wolfgang Kunath/Peter Schiske/Klaus Weiss/Friedrich Zemlin: Coma-Free Alignment of High-Resolution Electron Microscopes with the Aid of Optical Diffractograms Ultramicroscopy 3 (1978), p. 49–60.

[Hildebrandt, *Röntgenstrahlen*] Hildebrandt, Gerhard: Gekrümmte Röntgenstrahlen im schwach verformten Kristallgitter, Berlin 1958.

[Hosemann, *Lattice*] Hosemann, Rolf: Lattice Defects and Microparacrystals (This Week's Citation Classics), Current Contents (1990/12), p. 16.

[Hosemann, *Patent*] Deutsche Auslegeschrift Nr. 1033141. Verfahren und Vorrichtung zur Trocknung rißempfindlicher Ware, Anmelder: Dr. Rolf Hosemann, Anmeldetag: 25. November 1948, Bekanntmachung: 26. Juni 1958.

[Jacobi, Ranke *GaAs Surfaces*] Jacobi, Karl/Wolfgang Ranke: Structure and Reactivity of GaAs Surfaces, Progress in Surface Science 10 (1981), p. 1–52.

[Jander, Pfundt, *Leitfähigkeitsreaktionen*] Jander, Gerhart/Otto Pfundt: Leitfähigkeitsreaktionen und Leitfähigkeitsmessungen. Visuelle und akustische Methoden. Mit Beispielen für die Anwendung im Laboratorium und im Betrieb, (Die chemische Analyse 26), Stuttgart 1934.

[Jander, *Maßanalyse*] Jander, Gerhart: Maßanalyse. Theorie und Praxis der klassischen und elektrochemischen Titrierverfahren. I. und II. Teil, Berlin/Leipzig 1935.

[Kallmann, Mark, *Comptoneffekte*] Kallmann, Hartmut/Hermann Mark: Zur Größe und Winkelabhängigkeit des Comptoneffekte, Die Naturwissenschaften 13 (1925), p. 297–298.

[Kambe, *Cellular Method*] Kyozaburo Kambe: Theory of Low-Energy Electron Diffraction. II. Cellular Method for Complex Monolayers and Multilayers, Zeitschrift für Naturforschung 23 A (1968), p. 1280–1294.

[Karge, Ladebeck *Mordenite*] Karge, Hellmut G./Jürgen Ladebeck: Deactivation of Mordenite Catalysts on Reaction of Ethylene and Ethylbenzene, in: Proceedings of the 6th Canadian Symposium on Catalysis, Aug. 19–21, 1979. Ottawa, Ontario 1979, p. 140–148.

[Karl, *Hochdruckdilatometer*] Karl, Veit-Holger: Hochdruckdilatometer zum Bestimmen wichtiger Stoffeigenschaften für die Verarbeitung, Kunststoffe 68 (1978), p. 247–250.

[Klein, Überreiter, *Das Dilatometer*] Klein, Karl/Kurt Überreiter: Das Dilatometer als Hilfsmittel der Kunststoffforschung, Chemische Technik 15 (1942), p. 5.

[Klipping, *Hexamethylentetramin*] Klipping, Gustav: Messung des Dampfdruckes von Hexamethylentetramin und Beobachtung seiner thermischen Zersetzung, Diss. TU Berlin, Berlin 1954.

[Klipping, Lemke, Römisch, *Telescope*] Klipping, Gustav/Dietrich Lemke/Norbert Römisch: Liquid Helium Cooled Infrared Telescope for Astronomical and Atmospherical Measurements from Spacelab, Advances in Cryogenic Engineering 23 (1978), p. 628–633.

[Knipping, *Registrierapparat*] Knipping, Paul: Registrierapparat zur automatischen Aufnahme von Ionisierungs- und anderen Kurven, Zeitschrift für Instrumentenkunde 43 (1923), p. 241–256.

[Knipping, *Zehn Jahre*.] Knipping, Paul: Zehn Jahre Röntgenspektroskopie, Die Naturwissenschaften 10 (1922), p. 366–369.

[Kopfermann, *Kernmomente*] Kopfermann, Hans: Kernmomente, Leipzig 1940.

[Kratky, Sekora, Weber, *Kleinwinkelinterferenzen*] Kratky, Otto/Aurelie Sekora/ Hans Hermann Weber: Neue Kleinwinkelinterferenzen bei Myosin, Die Naturwissenschaften 31 (1943), p. 91.

[Kratky, Sekora, Röntgenstrahlen] Kratky, Otto/Aurelie Sekora: Bestimmung von Form und Größe gelöster Teilchen aus den unter kleinsten Winkeln diffus abgebeugten Röntgenstrahlen, Die Naturwissenschaften 31 (1943), p. 46–47.

[Ladenburg, Kopfermann, *Negative Dispersion*] Ladenburg, Rudolf/Hans Kopfermann: Experimenteller Nachweis der ,negativen Dispersion', Zeitschrift für Physikalische Chemie A139 (1928), p. 375–385.

[Langmuir, *Constitution*] Langmuir, Irving: The Constitution and Fundamental Properties of Solids and Liquids. Part I. Solids, Journal of the American Chemical Society 38 (1916), p. 2221–2295.

[Laue, *Röntgenstrahl-Interferenzen*] Laue, Max v.: Röntgenstrahl-Interferenzen, 3. neubearb. und erw. Aufl., Frankfurt am Main 1960.

[Laue, *Supraleitung*] Laue, Max v.: Theorie der Supraleitung, 1. und 2. Aufl., Berlin u.a. 1947 und 1949.

[Lehrecke, Lösungen] Lehrecke, Hans: Über die Bestimmung des Goldes in äußerst verdünnten Lösungen, Dissertation, Berlin 1922.

[London, Polanyi, *Adsorptionskräfte*] London, Fritz/Polanyi, Michael: Über die atomtheoretische Deutung der Adsorptionskräfte, Die Naturwissenschaften 18 (1930), p. 1099–1100.

[MPG, *FHI I*] A. Bradshaw, G. Ertl, H.-J. Freund, M. Scheffler and R. Schlögl, eds., *Fritz Haber Institute of the Max Planck Society Berlin* (Rosenheim: Format-Druck GmbH & Co., 1999).

[Müller, Ruska, *Durchstrahlungs-Elektronenmikroskop*] Müller, Käthe/Ernst Ruska: Über ein einfaches und leistungsfähiges permanentmagnetisches Durchstrahlungs-Elektronenmikroskop, Mikroskopie 23 (1968), p. 197–219.

[Nathansohn, *Rohstoffe*] Nathansohn, Alexander: Über die Technische Verarbeitung bleihaltiger Rohstoffe auf dem Wege über Bleitetrachlorid, Zeitschrift für Elektrochemie 13/14 (1922), p. 310–313.

[Niehrs, *Ausbreitung*] Niehrs, Heinz: Die Ausbreitung von Elektronenstrahlen in Kristallgittern, Berlin 1958.

[Pettinger, *Tunnelprozesse*] Pettinger, Bruno: Tunnelprozesse bei der Elektronenübertragung am Kontakt zwischen hochdotiertem Zinkoxyd und Elektrolytlösungen, München (Diss. TU) 1972.

[Plieth, Ruban, Smolczyk, *Laueit*] Plieth, Karl/Gerhard Ruban, Heinz-Günter Smolczyk: Zur Kristallstruktur des Laueits. $Fe_2Mn[PO_4/OH]_2 \times 8H_2O$, Acta Crystallographica 19 (1965), p. 485.

[Polanyi, Eyring, *Gasreaktionen*] Polanyi, Michael/Henry Eyring: Über einfache Gasreaktionen, Zeitschrift Für Physikalische Chemie 12 (1931), p. 279–311.

[Polanyi, Wigner, *Molekülen*] Polanyi, Michael/Eugene Wigner: Bildung und Zerfall von Molekülen, Zeitschrift für Physik 33 (1925), p. 429–434.

[Polnayi, *Problem*] Polanyi, Michael: Zum Problem der Reaktionsgeschwindigkeit, Zeitschrift für Elektrochemie 26 (1920), p. 161–171.

[Preisberg, *Simulation*] Preisberg, W.: Über die Möglichkeit der Simulation elektronenoptischer Systeme mit Hilfe digitaler Rechenanlagen, in: Daria Steve Bocciarelli (Eds.): Pre-Congress Abstracts of Papers Presented at the Fourth European Regional Conference; held in Rome, September 1 – 7, 1968, Rom 1968.

[Proske, Winkel, *Über die elektrolytische Reduktion*] Proske, Gerhard/August Winkel: Über die elektrolytische Reduktion organischer Verbindungen an der Quecksilber-Tropfelektrode. I. und II. Mitteilung, Berichte der Deutschen Chemischen Gesellschaft 69 (1936), p. 693–705 und p. 1917–1929.

[Rabinowitsch, *Mikroanalyse*] Rabinowitsch, Isaak: Ueber die Anwendung der Zentrifuge in der quantitativen Analyse, insbesondere der Mikroanalyse des Goldes, Dissertation, Berlin 1928.

[Roth, Ziehl, *Bombe*] Roth, Walther A./Ludwig Ziehl: Eine kalorimetrische Bombe mit Fenster zur Beobachtung des Verbrennungsvorganges, Zeitschrift für Elektrochemie 52 (1948), p. 219–220.

[Ruska, *Foundations*] Ruska, Ernst: Vibration Isolating Foundations for Electron Microscopes, in: David J. Goodchild/V. Sanders (Eds.): Electron Microscopy 1974. Abstracts of Papers Presented to the Eighth International Congress on Electron Microscopy. Held in Canberra, Australia. August 25–31 1974, Bd. 1., Canberra 1974, p. 24–25.

[Sass, *Photoemission*] Sass, Jürgen-Kurt: In Situ Photoemission, Surface Science 101 (1980), p. 507–517.

[Schalek, Szegvari, *Eisenoxydgallerten*] Schalek, Emmy/Andor Szegvari: Über Eisenoxydgallerten. Vorläufige Mitteilung, Kolloid Zeitschrift 32 (1923), p. 318–319.

[Schalek, *Koagulation*] Schalek Emmy/Andor Szegvary: Die langsame Koagulation konzentrierter Eisenoxydsole zu reversiblen Gallerten, Kolloid Zeitschrift 33 (1923), p. 326–334.

[Schmid, *Gold*] Schmid, Fritz: Über die Bestimmung des Goldes im Meerwasser. Doktorarbeit. Berlin 1922.

[Schmidt, *Massenspektrometrie*] Werner A. Schmidt: Massenspektrometrische Untersuchungen zur NH_3-Synthese an Spitzen aus Eisen, Angewandte Chemie 80 (1968), p. 151–152.

[Schoon, Klette, *Der Aufbau*] Schoon, Theodor/Hermann Klette: Aufbau typischer Adsorbentien, Die Naturwissenschaften 29 (1941), p. 652–653.

[Schoon, Thiessen, *Elektronen-Beugungsgerät*] Schoon, Theodor/Thiessen, Peter Adolf: Ein handliches Elektronen-Beugungsgerät und seine Anwendung zur Bestimmung des inneren Potentials von Ionenkristallgittern, Zeitschrift für Physikalische Chemie B 36 (1937), p. 195–215.

[Song, *photonenstimulierte Felddesorption*] Song, Sun-Dal: Aufbau und Test einer Anordnung zur photonenstimulierten Felddesorption mittels Excimerlaser und Flugzeitmassenspektrometrie, Berlin (Diss. TU) 1992.

[Stauff, *Mizellenarten*] Stauff, Joachim: Die Mizellenarten wässeriger Seifenlösungen, Kolloid Zeitschrift 89 (1939), p. 224–233.

[Szegevari, Wigner, *Stäbchensolen*] Szegvari, Andor/Eugene Wigner: Über elektrische Erscheinungen bei Stäbchensolen, Kolloidzeitschrift 33 (1923), p. 218–222.

[Thiessen, Wittstadt, *Änderung*] Thiessen, Peter Adolf/Werner Wittstadt: Erzwungene und spontane Änderung der molekularen Ordnung im gedehnten Kautschuk, Zeitschrift für Physikalische Chemie B 41 (1938), p. 33–58.

[Thiessen, *Seifen als Kolloide*] Thiessen, Peter Adolf: Seifen als Kolloide, Fette und Seifen 43 (1936), p. 149–152.

[Tributsch, *Desintegration*] Tributsch, Helmut: The Oxidative Desintegration of Sulfide Crystals by Thiobacillus ferrooxidans, Die Naturwissenschaften 63 (1976), p. 88.

[Überreiter, *Kautschuk und Kunstharze*] Überreiter, Kurt: Kautschuk und Kunstharze als ‚Flüssigkeiten mit fixierter Struktur', Kunststoffe 30 (1940), p. 170–172.

[Überreiter, Yamaura, *Surface Tension*] Überreiter, Kurt/Kazuo Yamaura: Dependence of Surface Tension and Polymer Concentration in Surface on Solvent Interaction of Polystyrene Solutions, Colloid and Polymer Science 255 (1977), p. 1178–1180.

[Überreiter, *Weichmachung*] Überreiter, Kurt: Über innere und äußere Weichmachung von makromolekularen Stoffen, Angewandte Chemie 53 (1940), p. 247–250.

[van Heel, *Classification*] van Heel, Marin: Classification of Electron Microscopical Images of Randomly Oriented Bio-Macromolecules, in: G.W. Bailey (Ed.): "Proc. 41st Ann. Meeting EMSA", San Francisco 1983, p. 762–763.

[Vetter, *Korrosion*] Vetter, Klaus J.: Die Korrosion des passiven Eisens in saurer Lösung, Zeitschrift für Elektrochemie 59 (1955), p. 67–72.

[Warrikhoff, *Dosimetrie*] Warrikhoff, Harald: Röntgenelemente für die Dosimetrie, Teil 1–3, Zeitschrift für angewandte Physik 18 (1964) p. 44–53 und p. 89–105.

[Wigner, *Erhaltungssätze*] Wigner, Eugene: Über die Erhaltungssätze in der Quantenmechanik Nachrichten der Gesellschaft der Wissenschaften zu Göttingen, Math.-Phys. Klasse (1927), p. 375–381.

[Witzmann, *Elementarvorgänge*] Witzmann, Hans: Elementarvorgänge bei Staub- und Nebelfiltration, Zeitschrift für Elektrochemie und angewandte physikalische Chemie 46 (1940), p. 313–321.

[Witzmann, *Mikrokolorimeter*] Witzmann, Hans: Ein Mikrocolorimeter mit Selensperrschichtzellen, Chemische Fabrik 12 (1939), p. 332–334.

[Zeitler, *Electron Tomography*] Zeitler, Elmar: Electron Tomography. Three Dimensional Imaging with the Transmission Electron Microscope, New York 1992, p. 63–89.

[Zemlin, *Procedure for Alignment*] Zemlin, Friedrich: A Practical Procedure for Alignment of a High Resolution Electron Microscope, Ultramicroscopy 4 (1979), p. 241–245.

Archives

ARHG   Archiv der Robert-Havemann-Gesellschaft, Berlin.
BMA   Bundesmilitärarchiv, Freiburg im Breisgau.
BNA   British National Archive, Kew, England.
GSTA   Geheimes Staatsarchiv Preußischer Kulturbesitz, Berlin.
HHP   Harold Hartley Papers, Churchill College Cambridge.
MPGA   Archiv der Max-Planck-Gesellschaft, Berlin-Dahlem.
PHP   Paul Harteck Papers, Rensselaer Polytechnic Institute, Troy, New York.
RAC   Rockefeller Archive Center, Sleepy Hollow, New York.

List of Figures

MPI für Mikrostrukturphysik, Halle: Box: Chap. 5, MPG and the German Reunification.

Archiv der Physikalisch-Technischen Bundesanstalt: Fig. 1.1.

Bildarchiv Preußischer Kulturbesitz: Fig. 1.2.

Reichsgesetzblatt, Teil I, Nr. 34, 7 April 1933, p. 1975: Box: Chap. 3, Berufs-
beamtengesetz.

Robert-Havemann-Gesellschaft: Box: Chap. 4, Robert Havemann.

Courtesy of Kenji Tamaru: Box: Chap. 2, Haber and Science Policy.

University of Utah, Special Collections: Fig. 2.22.

With special thanks to David Vandermeulen: Box: Chap. 1, Leopold Koppel.

http://en.wikipedia.org/wiki/File:Heisenberg,W._Wigner,E._1928.jpg
(Friedrich Hund Nachlass): Fig. 2.26.

„Die Woche" Berlin 1910, p. 1778: Fig. 1.3.

Kurt Zierold: Forschungsförderung in 3 Epochen, Wiesbaden 1968, Tafel 6., p. 113:
Fig. 2.12.

Index

Jeremiah James (*1975) received his Ph.D. in the history of science from Harvard University in 2008 with a dissertation entitled *Naturalizing the Chemical Bond: Discipline and Creativity in the Pauling Program.* As a graduate student he was a Pre-Doctoral Fellow at the MPI for the History of Science and at the former Dibner Institute in Cambridge, Massachusetts, as well as an Edelstein Student at the Chemical Heritage Foundation in Philadelphia. He joined the Centennial Project in 2008 as a Post-Doctoral Fellow, and during his time in Dahlem has also worked closely with the Project on the History and Foundation of Quantum Physics at the MPI for the History of Science.

Thomas Steinhauser (*1964) studied chemistry, Italian, and history of science at the University of Regensburg, where he worked on Galileo and the Liebig-Wöhler correspondence. The topic of his PhD thesis was the history of nuclear magnetic resonance (NMR). In 2008 he researched the construction and diffusion of the first self-recording infrared spectrophotometers as a Scholar-in-Residence at the Deutsches Museum, Munich. Subsequently, he was a member of the Centennial Group until 2011. Currently he is working at the University of Bielefeld on chemistry experts and threshold limit values of chemical substances. His additional research interests focus on spectroscopic instrumentation in 20th century chemistry.